Geosynthetic Encased Columns
for Soft Soil Improvement

Geosynthetic Encased Columns for Soft Soil Improvement

Marcio de Souza Soares de Almeida
Institute for Graduate Studies and Research in Engineering (COPPE), Federal University of Rio de Janeiro (UFRJ), Brazil

Mario Vicente Riccio Filho
Graduate School of Engineering (PEC-UFJF), Federal University of Juiz de Fora (UFJF), Brazil

Iman Hosseinpour Babaei
Department of Civil Engineering, The University of Guilan, Iran

Dimiter Alexiew
Consultant Geosynthetics & Geotechnics, Germany

CRC Press
Taylor & Francis Group
Boca Raton London New York

CRC Press is an imprint of the
Taylor & Francis Group, an **informa** business

A BALKEMA BOOK

CRC Press
Taylor & Francis Group
6000 Broken Sound Parkway NW, Suite 300
Boca Raton, FL 33487-2742

First issued in paperback 2020

ISBN-13: 978-1-138-03878-3 (hbk)
ISBN-13: 978-0-367-78076-0 (pbk)

Library of Congress Cataloging-in-Publication Data
Names: Almeida, Marcio de Souza Soares de, author. | Riccio Filho, Mario Vicente, author. | Babaei, Iman Hosseinpour, author. | Alexiew, Dimiter, author.
Title: Geosynthetic encased columns for soft soil improvement / Marcio de Souza Soares de Almeida, Graduate School of Engineering (COPPE), Federal University of Rio de Janeiro (UFRJ), Brazil, Mario Vicente Riccio Filho, Faculty of Engineering, Federal University of Juiz de Fora (UFJF), Brazil, Iman Hosseinpour Babaei, Department of Civil Engineering, The University of Guilan, Iran, Dimiter Alexiew, Dr-Ing., Independent Geosynthetics & Geotechnical Consultant, Former Head of Technical Department at Huesker Ltd, Germany.
Description: London ; Boca Raton : CRC Press/Balkema is an imprint of the Taylor & Francis Group, an Informa Business, [2019] | Includes bibliographical references and index.
Identifiers: LCCN 2018039217 (print) | LCCN 2018040595 (ebook) | ISBN 9781315177144 (ebook) | ISBN 9781138038783 (hbk)
Subjects: LCSH: Geosynthetic encased columns. | Reinforced soils. | Piling (Civil engineering) | Embankments—Materials. | Foundations—Materials.
Classification: LCC TA788 (ebook) | LCC TA788 .A46 2019 (print) | DDC 624.1/54—dc23
LC record available at HYPERLINK "https://protect-us.mimecast.com/s/8cAmCNkE8qi0oMJ rZIR9Yf8?domain=lccn.loc.gov" https://lccn.loc.gov/2018039217

Visit the Taylor & Francis Web site at
http://www.taylorandfrancis.com

and the CRC Press Web site at
http://www.crcpress.com

Contents

Foreword

Foundations on very soft soils are always problematic. However, when the undrained shear strength is below some 15 kPa, even solutions such as stone columns prove inadequate. This is because the horizontal support provided by the soft soil should be able to resist the horizontal outward pressures in the column. Nonetheless, the ingenuity of designers and contractors and the ability of geosynthetic producers to tailor the material properties to satisfy design needs led to the use of high-strength geotextiles to construct geosynthetic encased columns (GECs). In this way, geosynthetics were able to provide yet another innovative solution to a geotechnical problem: foundation design, in this case, a sub-discipline that is not generally associated with the use of geosynthetics.

GECs may serve as foundation elements in very soft soils, including underconsolidated clays, peats, and sludge. Since their advent in the 1990s, a significant number of successful projects have been completed in countries such as Germany, Sweden, Holland, Poland, Brazil, and Turkey.

Geosynthetic Encased Columns for Soft Soil Improvement is a response to the increasing demand for engineers knowledgeable in two of the most unique materials that could be faced in geotechnical projects: geosynthetics and soft soils. In writing this book, the authors focused the presentation of the information on the different phases that may be involved in a GEC project: from pre-design using charts to design using analytical methods to the lessons learned from instrumentation and numerical simulations. The authors' comprehensive treatment of design approaches, construction options, and case studies not only facilitates an understanding of the principles of GEC technology, but also provides resources for actual implementation of this comparatively new method of soft soil improvement.

Jorge G. Zornberg, Ph.D., P.E., F. ASCE
Professor, Department of Civil Architectural and Environmental Engineering
The University of Texas at Austin
Immediate past-President
International Geosynthetics Society (IGS)

Preface

The geosynthetic encased column (GEC) is a relatively recent method developed for soft soil improvement. The method was first introduced as a concept in the 1980s, and the first practical applications started in the 1990s. GECs have been widely used in some parts of the world for the last three decades. However, there is not yet in the literature a book summarizing the knowledge accumulated during this period in relation to this soft ground improvement technique.

The purpose of this book is to provide readers with the GEC fundamentals and practical applications. Chapter 1 presents the general principles of this ground improvement technique, including the methods used for GEC installation and how the material properties may be selected. Chapter 2 presents design methods; thus, settlement calculations by means of analytical methods and stability calculations by limit equilibrium methods are explained in detail. Chapter 3 presents calculation examples illustrating the usual steps to be done for both service limit state and ultimate limit state designs. Then, field performances exemplifying practical applications of the GEC technique are presented in Chapter 4 for some case histories. Numerical analyses, often used in design to complement analytical methods, are presented in Chapter 5. Annexes I and II at the end contain the charts developed to perform settlement calculations.

The book combines the experiences of four authors with different academic and industry backgrounds to describe GEC design and performance. The book is aimed at civil engineers in general, particularly geotechnical engineers, working either in design or in practice, at graduate students, and at senior undergraduate students.

The authors would like to acknowledge all who have added important comments and suggestions to this book, namely Prof. Marcos Futai, Prof. Chungsik Yoo, Eng. Cristina Schmidt, Dr. Marina Miranda, and Dr. Sunil Mohapatra. The schematic drawings were prepared by Eng. Rafael Lima Carvalho whom is appreciated. The authors also thank Juliana Rigolon for her assistance in the preparation of the pre-design charts.

The authors would like to thanks HUESKER Synthetics for the general support and to allow the use of its photos in the cover page.

About the authors

Marcio de Souza Soares de Almeida earned his Civil Engineering degree at the Federal University of Rio de Janeiro in 1974 and obtained his M.Sc. at COPPE/UFRJ in 1977 when he joined COPPE. Marcio got his Ph.D. from the University of Cambridge, UK, in 1984. Then he returned to UFRJ and in 1994 became Professor of Geotechnical Engineering. His postdoc was in Italy (ISMES) and at NGI, Norway, in the early 1990s. He has published around 300 technical publications including more than 70 articles in international refereed journals and has supervised around 90 doctoral and master dissertations. He received the Terzaghi and José Machado awards from the Brazilian Association of Geotechnical Engineering (ABMS). He has served on the editorial board of *Geotechnique*, *Geotechnique Letters*, *ICE-Ground Improvement*, and *International Journal of Physical Modelling*, among others. In June 2015, he delivered the "Coulomb Lecture" at the French Geotechnical Society. His main topics of interest are soft clay engineering, ground improvement, marine geotechnics, and physical modelling.

Mario Vicente Riccio Filho has been a Civil Engineer at the Federal University of Juiz de Fora (UFJF) since 1998, obtaining his M.Sc. in 2001 and his Ph.D. in 2007 at COPPE/ Federal University of Rio de Janeiro. He has worked for five years for geotechnical engineering companies. His postdoc was done at COPPE/Federal University of Rio de Janeiro, finishing in 2015. In 2008, he was awarded the Dirceu de Alencar Velloso prize promoted by AEERJ (Rio de Janeiro Association of Engineering Companies). Also, in 2017, he received the Mokshagundam Visvesvaraya prize by ICE (Institution of Civil Engineers, England). Since 2015, he has been Associate Professor of Civil Engineering at the Federal University of Juiz de Fora. His main research interests are soft soil improvement, geosynthetics, and soil retaining systems.

Iman Hosseinpour Babaei received his B.Sc. in Civil Engineering and M.Sc. in Geotechnical Engineering from the Noshirvani University of Technology, Iran, in 2005 and 2008, respectively. He then completed his Ph.D. in Geotechnical Engineering with the Graduate School of Engineering (COPPE) at Federal University of Rio de Janeiro (UFRJ), Brazil, in 2015. Following a two-year postdoc at COPPE/UFRJ, he joined the Department of Civil Engineering at the University of Guilan, Iran, as an assistant professor. He was given the Brazilian Association Geotechnical Engineering's (ABMS) excellent doctoral thesis award in 2016. He is also the recipient of 2017 Institution of Civil Engineers (ICE, UK) award for a paper published in the *Ground Improvement Journal*. His research interests involve soft soils, geosynthetics, large-scale experimental tests, and numerical analyses.

Dimiter Alexiew earned his Dipl.-Ing. in Civil Engineering at the Technical University (TUCE), Sofia, in 1978. Then he worked as researcher, design engineer, and consultant for geotechnics at the Bulgarian Institute for Construction, and was simultaneously honorary assistant professor for geotechnics at TUCE. He received his doctorate in civil engineering and geotechnics from TUCE in 1990. After working as design engineer and consultant at the Bavarian LGA Geotechnical Institute in Nuremberg, Germany, he was until 2017 Head of Engineering and Technical Director of Huesker Synthetic, Germany. Recently he founded his own independent consulting firm for geosynthetics and geotechnics. His activities include a wide range of geotechnical projects, many of them pioneering, development of design procedures and novel geosynthetics, and co-authoring of codes and standards. He has written about 230 technical publications, and has given more than 100 lectures and reports. He is invited lecturer at the University in Siegen, Germany, and member of several German and international societies and committees and IGS Council Member.

Symbols

a_E: area replacement ratio (-)
A_c: cross-sectional area of the column (m²)
A_E: column influence area (m²)
b_c: half of the width of the equivalent wall (m)
B: half of the equivalent width in 2D plane strain condition (m)
B: width of adjacent piles (m)
c': drained cohesion (kPa)
Cc: coefficient of compression (-)
c_r, c_h: coefficient of horizontal consolidation (m² s⁻¹)
c_{rm}: modified coefficient of radial consolidation (-)
Cs: coefficient of swelling (-)
c'_s: soft soil effective cohesion (kPa)
c_v: coefficient of vertical consolidation (m² s⁻¹)
C_a: coefficient of secondary compression (-)
d_c: diameter of the column (m)
d_e: diameter of the influence area (m)
e_0: initial void ratio (-)
E_c: elastic modulus of column material (kPa)
E_{oed}: constrained modulus of soft soil (kPa)
E_s: elastic modulus of soil (kPa)
E_u: undrained elastic modulus (kPa)
F: hoop force developed in geosynthetic encasement (kN m⁻¹)
FOS: safety factor for uncertainties in production and data extrapolation (-)
F_0: geosynthetic maximum tensile force in quick wide strip test (kN m⁻¹)
F_{all}: geosynthetic allowable tensile force (kN m⁻¹)
f_c: safety factor to decrease soil cohesion (-)
f_{fi}: safety factor to decrease soil angle of friction (-)
F'_m: consolidation function (-)
f_s: safety factor to increase the unit weight of soil (-)
f_t: safety factor for external permanent load (-)
f_q: safety factor for external live load (-)
G_s: specific gravity of the grains (-)
h_0: initial column length (m)
H_{cri}t: critical height of embankment (m)
H_d = longest drainage path due vertical flow (m)

H_{em}: embankment height (m)
H_s = soft soil thickness (m)
H_{wp}: thickness of working platform (m)
J: geosynthetic stiffness modulus (kN m^{-1})
J_{bs}: stiffness modulus of basal reinforcement (kN m^{-1})
J_{en}: stiffness modulus of the encasement (kN m^{-1})
k_h: horizontal/radial hydraulic conductivity (m s^{-1})
k_s = coefficient of permeability in the smeared zone (m s^{-1})
k_v: vertical hydraulic conductivity (m s^{-1})
K_0: coefficient of at-rest earth pressure (-)
K_0*: modified coefficient of at-rest earth pressure
K_a: coefficient of active earth pressure (-)
K_p: coefficient of active earth pressure (-)
L_{col}: column's length (m)
L_{enc}: length of encasement (m)
m: hyperbolic power law (-)
m_v = coefficient of volumetric compressibility (m^2 kN^{-1})
n: stress concentration factor (-)
n_g: Poisson's ratio of the geosynthetic (-)
p*: effective vertical stress in middle of soft soil layer (kPa)
p_{ref}: reference effective vertical stress (kPa)
P_h: lateral force acting on the pile (kN m^{-1})
q: vertical applied load (kN m^{-2})
r_c: column's radius (m)
r_{geo}: original encasement radius (m)
R: radius of the influence area (m)
RF_{amb}: reduction factor for chemical and environmental damages (-)
RF_{dm}: mechanical damage reduction factor (-)
RF_f: creep reduction factor (-)
RF_{joint}: reduction factor for joints/seems (-)
s: column center to center spacing (m)
S: settlement (m)
S_c: settlement on encased column (m)
S_{mr} = d_s/d_c = ratio between the smeared zone diameter (d_s) and the column's diameter (d_c) (-)
S_s: settlement on soft soil (m)
S_t: clay sensitivity (kPa)
S_u: undrained shear strength (kPa)
t: time (day)
T: geosynthetic tensile force (kN m^{-1})
T_{rm}: modified time factor (-)
U: degree of consolidation (-)
w_n: natural water content (%)
w_L: liquid limit (%)
w_p: plastic limit (%)
z: depth (m)
γ': submerged unit weight of soft soil (kN m^{-3})
γ'_c: column fill material submerged unit weight (kN m^{-3})

γ_n: natural unit weight of soil or gravel (kN m^{-3})

γ_{sat}: saturated unit weight of soil (kN m^{-3})

γ_w: specific unit weight of water (kN m^{-3})

$\delta_{h,max}$: maximum horizontal displacement (m)

$\delta_{v,max}$: maximum vertical displacement (m)

Δr_c: column radial/horizontal deformation (m)

Δ_{rgeo}: encasement radial/hoop deformation (m)

Δu: excess pore water pressure (kPa)

$\Delta\sigma_0$: total applied embankment load (kPa)

$\Delta\sigma_0$: total applied stress (kPa)

$\Delta\sigma_{3,geo}$: horizontal (radial) stress supported by geosynthetic (kPa)

$\Delta\sigma_{h,c}$: increase in horizontal stress on top of the column (kPa)

$\Delta\sigma_{h,diff}$: horizontal stress acting on the geosynthetic (kPa)

$\Delta\sigma_{h,geo}$: encasement horizontal stress (kPa)

$\Delta\sigma_{h,s}$: increase in horizontal stress on top of the surrounding soil (kPa)

$\Delta\sigma_{v,c}$: increase in vertical stress on top of the column (kPa)

$\Delta\sigma_{v,s}$: increase in vertical stress on top of the surrounding soil (kPa)

ε_r: radial/hoop strain in geotextile encasement (%)

$\sigma_0 = $ *in situ* vertical stress (kPa)

σ_1: major vertical stress (kPa)

σ_3: minor horizontal (radial) stress (kPa)

σ'_{vm}: over consolidation stress (kPa)

υ: Poisson's ratio (-)

ϕ': drained angle of friction ($^\circ$)

ϕ_c: friction angle of column material ($^\circ$)

ϕ'_s: soft soil effective friction angle ($^\circ$)

ϕ'_{sub}: equivalent friction angle of the encased column in plane strain condition ($^\circ$)

ψ_c: dilatancy angle of column material ($^\circ$)

Figures and charts

Figures

Charts

Tables

Chapter 1

Generalities

1.1 Introduction

The design and construction of the structures on soft soil deposits have been, historically, a challenge for the geotechnical engineers so that serviceability and limit state conditions as well as cost and time schedule are properly addressed. In order to meet these requirements, a great variety of construction methods is available. These may include, for instance, soft soil replacement, basal embankment reinforcement, use of lightweight fill material, prefabricated vertical drains and surcharge or vacuum to accelerate settlements, stage construction, use of granular stone columns, or different alternatives of soil cement mixture such as deep mixing or pile embankments with basal reinforcement (Almeida and Marques, 2013).

Among all available construction methods, the soft ground treatment with stone columns is one of the most widely adopted for reducing settlement and improving stability and load capacity (Poorooshasb and Meyerhof, 1997; Greenwood, 1970). However, when stone columns are installed in extremely soft soils, they may not provide significant load capacity owing to low lateral soil confinement. McKenna *et al.* (1975) reported cases in which the stone column was not restrained by the surrounding soft clay, which led to excessive bulging and soft clay squeezing into the voids of the stone aggregates, reducing the bearing capacity of the stone column as well as its drainage capacity.

The use of traditional stone columns is usually limited to the values of soft soil undrained strength S_u around 15 kPa (EBGEO, 2011), thus confining the stone column with a high-tensile-stiffness geosynthetic encasement (Raithel *et al.*, 2002; Alexiew *et al.*, 2005; Di Prisco *et al.*, 2006; Murugesan and Rajagopal, 2006) can overcome this difficulty. There are also limited reports (Wehr, 2006; Yee and Raju, 2007) that the soft soils with S_u values lower than 15 kPa have been treated with traditional stone columns.

Ghionna and Jamiolkowski (1981) and Van Impe and Silence (1986) were probably the first to recognize that columns could be encased by geotextiles. They produced an analytical design technique to assess the required geotextile tensile strength, thus an ultimate limit state (ULS) analysis. Details on this technique were also provided by Kempfert *et al.* (1997). In addition, Raithel and Kempfert (2000) produced an overall design calculation method for analyzing the service limit state; say assessing column and soft soil settlements based on the geotextile radial tensile stiffness. An update, including use on recent projects in Europe, was provided by Raithel *et al.* (2005); Alexiew *et al.* (2005); Alexiew and Thomson (2014); andAlexiew and Raithel (2015) and in South America by De Mello *et al.* (2008).

1.2 General principles

The general scheme of the bearing system with geosynthetic encased columns (GECs), developed in Germany in the mid-1990s, is depicted in Figure 1.1. The encasement used for

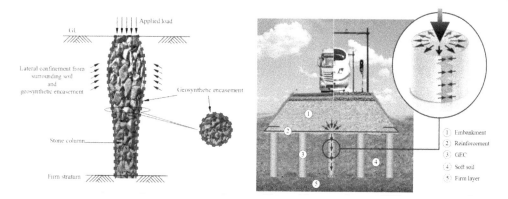

Figure 1.1 (a) Scheme of GEC (Murugesan and Rajagopal, 2006); (b) outline of an embankment on soft soil over GECS (Alexiew and Thomson, 2014)

the columns commonly consists of a woven geotextile with high-tensile modulus and low creep coefficient, which results in favorable drainage characteristics of the granular column and low strains in the geotextile. The column filling material can be sand or gravel; the latter, however, provides higher overall stiffness of the encased column, but has to be compatible with the geosynthetic material used to prevent its damage. In addition, the use of granular spoilage is possible in some cases. The geosynthetic encasement also controls the column diameter, minimizes material losses, increases overall column stiffness, and avoids granular column contamination, thus preserving the drainage features.

Due to the higher stiffness of the GECs, load concentration (arching) occurs, thus reducing the vertical stresses on the soft foundation soils. The vertical load on a GEC generates also horizontal radial normal stresses outwards and radial widening of the column. This consequently results in counter-pressure from the surrounding soft soils and a confining resistance from the encasement, the latter being the key difference from "conventional" stone columns. The mobilized confining ring tensile force T_{mob} in the encasement depends on its tensile stiffness (modulus J) and hoop strain. The tensile force T_{mob} and the corresponding radial strain (elongation) control the radial behavior and consequently the vertical performance of the GEC in terms of settlement and bearing capacity. The less the GEC compresses, the higher is the embankment load supported by the GEC and the smaller is the vertical stress taken by the soft soils in between GECs.

1.3 Applications (Alexiew and Thomson, 2014)

GEC may be applied in soft soil deposits with values of undrained strength S_u lower than 30 kN m^{-2}, being better suited for S_u values lower than 15 kPa. However, it is also possible to apply GEC in the case of S_u values as low as 5 kPa. The main range of interest of GEC are soft soils with constrained modulus E_{oed} between 0.5 and 3.0 MPa and their thickness from 8 to 30 m. The minimum recommended embankment height above the GECs is 1.5 m.

The range of settlement values compensated during the construction stage in GEC applications is usually 0.1–0.5 m. Moreover, as the GECs work as "mega-drains," primary consolidation and settlements occur relatively quickly in comparison with prefabricated vertical drains. Post-construction settlements in GEC applications are usually quite small, and differential settlements are inexistent when the embankment height is above the critical height.

GECs are quite suitable in the case of foundations sensitive to lateral thrust such as piles in the vicinity of high embankments, stock piles, or other directly founded loads. GECs have also been applied in seismic areas for keeping the integrity of granular columns under "shearing" seismic impact. A further application of GECs is for existing railway embankments being upgraded for higher speed trains, as GECs increase their dynamic stability (Alexiew et al., 2015).

1.4 Execution methods

Encased columns can be installed with or without lateral displacement of the soft soil, thus two different methods are generally available with regards to the GEC construction technology.

The first technique, as shown in Figure 1.2, is the displacement method, in which a closed-tip steel pipe is driven down into the soft soil followed by the insertion of the circular woven geotextile and sand or gravel fill in the sequence. The tip opens when the pipe is pulled upwards under optimized vibration designed to compact the column.

The displacement method is commonly used for very soft soils (e.g., $S_u < 15$ kN m^{-2}). GECs executed with the displacement of soft soil usually have a diameter of approximately 0.80 m, and the diameter of the geotextile is ideally equal to the inner tube diameter (Alexiew et al., 2003). The column spacing ranges typically between 1.5 and 2.5 m and the tensile stiffness modulus (J) of the geotextile generally varies between 1500 and 6500 kN m^{-1}.

The second construction technique is the replacement method with excavation of the soft soil inside the pipe. With the replacement method, an open steel shaft is driven deep into the bearing layer and the soil within the shaft is removed out by augering, as illustrated in Figure 1.3. The replacement method is preferred for soils with relatively higher penetration resistance, or when vibration effects on nearby buildings and road installation have to be minimized.

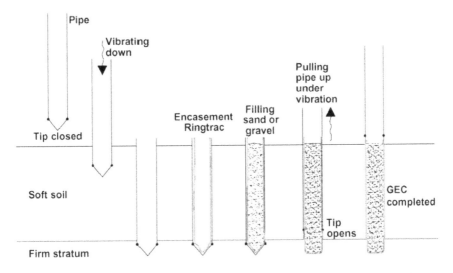

Figure 1.2 Displacement method for GEC installation (Alexiew et al., 2005, after Huesker)

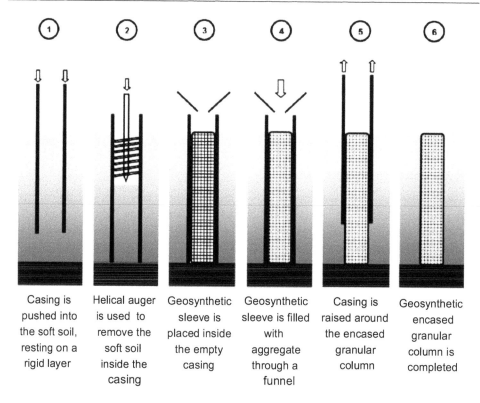

| Casing is pushed into the soft soil, resting on a rigid layer | Helical auger is used to remove the soft soil inside the casing | Geosynthetic sleeve is placed inside the empty casing | Geosynthetic sleeve is filled with aggregate through a funnel | Casing is raised around the encased granular column | Geosynthetic encased granular column is completed |

Figure 1.3 Replacement method for GEC execution (Gniel and Bouazza, 2010)

1.5 Material properties selection

1.5.1 Soft soil

The design of GECs-supported embankments requires high-quality soft soil parameters to be obtained in a well-specified and controlled site investigation program. This includes *in situ* and laboratory tests performed on good quality undisturbed soft soil samples. Among the soft soil parameters, the compressibility characteristics, the stress history, and the strength properties are those that mainly control the settlement and stability of the embankment over soft soil deposit.

The Standard Penetration Test (SPT) is the dominant *in situ* test for preliminary soil investigation, but very often, it is complemented by other *in situ* and laboratory tests. The Vane Shear Test (VST) is usually employed to determine the *in situ* undrained strength and clay sensitivity. The Piezocone Test (CPTu), with pore pressure measurements, is particularly effective for soft clays, as it allows the estimation of both strength and consolidation characteristics, which are key properties of such soft soils (Lunne *et al.*, 1997; Schnaid, 2005; Robertson and Cabal, 2015). In addition, the CPTu provides the soil stratigraphy, as well as an estimation of the stress history.

Table 1.1 summarizes the tests usually performed and the soft soil parameters obtained from each test. The parameters shown in Table 1.1 are defined in the list of symbols.

Table 1.1 Recommended laboratory and in situ tests and geotechnical design parameters (Almeida and Marques, 2013)

Test	Type	Aim of the test	Main soil parameters	Other parameters	Notes and recommendations
Laboratory	Complete characterization	General characterization of the soil; interpretation of other tests	w_n, w_L, w_P, G_s, grain size distribution curves	–	Recommended to access the organic matter in very organic soils and peat
	Consolidation test	Calculation of settlements and settlements vs. time curve	C_c, C_s, σ'_{vm}, c_v, e_o, E_{oed}	E_{oed}, c_α	Essential for calculating the magnitude and rate of settlements; can be replaced by CRS test
	Triaxial CU	Stability calculations; parameters for deformability calculations 2D or 3D (FEM)	S_u, c', ϕ'	E_u	CAU test (anisotropic consolidation) is more indicated. Pore pressure measured.
In situ	Standard Penetration Tests (SPT)	First test to be performed; used to specify the remaining tests	Soil layer description	Water content can be measured at low cost-benefit ratio	Measurement of water content requires special procedures to provide meaningful data
	Vane	Stability calculations	S_u, S_t	–	Essential for determining the clay undrained strength
	Piezocone (CPTu)	Stratigraphy; settlements vs. time (from dissipation test)	Soil layering profile, S_u profile, c_h (c_v)	Soil behavior, OCR profile, K_0, E_{oed}	–

Cc: compression index; Cs: swelling index; c_v: coefficient of vertical consolidation; c_h: coefficient of horizontal consolidation; c': drained cohesion; ϕ': drained angle of friction; E_{oed}: constrained modulus; e_o: initial void ratio; K_0: at-rest coefficient of thrust; OCR: over consolidation ratio; S_u: undrained shear strength; σ'_{vm}: pre-consolidation stress; E_u: undrained elastic modulus; S_t: clay sensitivity; w_n: natural water content; w_L: liquid limit; w_P: plastic limit; G_s: specific gravity of the grains; c_α: coefficient of secondary compression.

1.5.2 Granular column material

The geotechnical properties of the granular column fill material can be determined by common laboratory tests such as direct shear or triaxial tests, thus allowing to obtain the parameters to be used in either numerical or analytical calculations. These tests, however, are performed very often.

For GEC executed with stone columns, the gravel should be clean, preferably crushed stone, hard, and free from organics or other deleterious materials, and its degradation using the Los Angeles testing machine should be less than 45% loss and its grain size should be between 12 and 75 mm (Castro, 2017).

The relative density of the gravel in the stone columns is not usually measured and may vary along the length of the column, in a similar manner as the column diameter. A proper stone column construction should achieve relative densities of the gravel above 75% (Barksdale and Bachus, 1983). The friction angle of the columns (ϕ_c) has a notable influence on the results of a stone column treatment and limits up to 50°.

The Young's modulus of the columns is usually between 25 and 100 MPa and varies with the granular material used and the confining pressure. A hyperbolic power law "m" is sometimes used to reproduce the stress-dependent stiffness of the granular columns (normally around 0.3). Table 1.2 lists the material properties reported in the literatures to model the behavior of the encased columns in the numerical analyses (Castro, 2017; Hosseinpour, 2015).

1.5.3 Geosynthetic encasement

The geosynthetic encasement can be modeled as a flexible membrane not supporting compressive stresses with a negligible thickness, which behaves as a linear elastic material with a modulus of J_g ranging from 1500 to 6000 kN m^{-1}.

The tensile strength is usually reached for circumferential strains of around 5–10%, which implies strength values of 100–300 kN m^{-1} (Alexiew and Raithel, 2015). It is common

Table 1.2 Parameters to model the encased column in the numerical analyses

Reference	ϕ_c (°)	ψ_c (°)	E_c (kPa)	υ (−)	γ_{sat} (kN m^{-3})
Malarvizhi and Ilamparuthi (2007)	48	4	9	0.3	16
Yoo (2010)	40	10	40	0.3	23
Almeida et al. (2013)	38	50	15.5	0.3	18
Hosseinpour et al. (2015)	40	–	80	–	20
Alexiew and Raithel (2015)	32.5	0	–	–	20
Khabbazian et al. (2015)	35	0	30	0.2	–
Chen et al. (2015)	38	10	40	0.3	22
Castro (2017)	32.5	5	50	0.33	20
Ayadat and Hanna (2005)	40	5	76.5	0.3	19
Murugesan and Rajagopal (2006)	40	5	85	0.3	18.5
Lo et al. (2010)	38	0	85	0.33	20
Keykhosropur et al. (2012)	32.5	2.5	50	0.3	19

E_c: column elastic modulus; γ_{sat}: saturated unit weight; υ: Poisson's ratio; ϕ_c: column friction angle; ψ_c: dilation angle.

for geosynthetics to be anisotropic, and then different properties should be input for each direction. Some other geosynthetic features, such as creep and damage during installation, are usually considered through reduction factors. In the recent numerical analyses (Liu *et al.*, 2017) in which the geosynthetic material was modeled as a continuum element of small thickness, it was found that it is necessary to ensure that it does not support compressive stresses, and hence, bending moments. Little attention is usually paid to the Poisson's ratio of the geosynthetic (n_g), but it may have important consequences on the results (Castro, 2017). Common geosynthetics for column encasement are woven geotextiles. For woven geotextiles, the two directions work nearly independently; so, it seems logical to use values of the Poisson's ratio close to zero. Soderman and Giroud (1995) propose values of $n_g = 0.1$ for woven geotextiles and $n_g = 0.35$ for non-woven geotextiles.

Chapter 2

Design methods

2.1 Overview

This chapter is focused on the design and analysis of the geosynthetic encased columns (GECs). As in any geotechnical structure, the design procedure has to consider two main aspects:

a. Serviceability Limit State (SLS) design or "vertical" design: define the GEC layout and characteristics to meet the allowed settlement criteria. This criterion also includes the verification of encasement tension failure (ULS design);

b. Ultimate Limit State (ULS) design or "global" stability design: assessment of the global failure of the embankment-GECs-soft soil system.

In this chapter, an analytical method used for the vertical design and analysis of the GECs-treated soft foundation is introduced. An analytical solution for prediction of the time-dependent settlement is also presented. In addition, and for the purpose of global stability, stability analysis of embankments over GECs is also explored.

2.2 Vertical design: "general principles"

The most commonly used method for the design of GEC systems was proposed by Raithel (1999), which later was complemented by Raithel and Kempfert (2000). Other calculation methods were subsequently proposed by many researchers (e.g., Castro and Sagaseta, 2010; Zhang et al., 2011; Pulko et al., 2011; Castro and Sagaseta, 2013; Zhang and Zhao, 2015), which are also summarized briefly in this chapter. Raithel and Kempfert's (2000) method is derived from the solution presented by Ghionna and Jamiolkowski (1981), which was the first one developed for the analysis of stone columns' improved foundation. Later, Van Impe and Silence (1986) presented a methodology for this type of calculation including an Ultimate Limit State analysis without considering the encasement strains or the deformations and settlements. Raithel and Kempfert's (2000) method complements both the methodologies presented in 1981 and 1986 considering the geosynthetic strains and settlements. This developed procedure was then implemented in EBGEO's (2011) recommendation with some small modifications. It is probably nowadays the most-used solution for vertical design and analysis of GEC systems.

The Raithel and Kempfert (2000) method utilizes the unit cell concept, in which each column is responsible for the equilibrium of a portion of soil surrounding a single encased column. The designed values provided through iterations are (a) tensile force in the geo-synthetic encasement; (b) settlement separately on the soil and encased column; (c) specific circumferential strain in the geosynthetic encasement; (d) vertical stress on the top of the encased column; and (e) vertical stress on the top of the soft soil. This method considers the following hypotheses through the calculation process:

- The settlements occurring in the firm layer (where the columns rest) are disregarded;
- The settlements of the column and the surrounding soft soil are equal;
- The columns mobilize the active state condition, thus the coefficient of active thrust, K_{ac}, is applied;
- When the columns are installed using the replacement method, the at-rest coefficient of thrust is applied to the soil $K_s = K_{0,s} = 1 - \sin \phi'$; if the displacement method is used, the lateral thrust coefficient is increased in relation to the at-rest coefficient of thrust, $K_s = K^*_{0,s}$. Both coefficients are relative to the instant before embankment construction;
- The geosynthetic encasement has a linear-elastic behavior;
- Calculations consider the soft soil long-term drained behavior (thus, effective stress parameters apply) because the maximum settlement and tensile force in the encasement occur for this condition.

It is observed that the use of increased coefficient of thrust $K^*_{0,s}$ (2–3 times $K_{0,s}$) can result in zero tension values on the geosynthetic encasement. It means that the soil suitably confines the columns, which then no longer require the restriction promoted by the encasement. This soil-confining effect of $K^*_{0,s}$, however, may cease over time due to the phenomenon of stress relaxation occurring in the soft clay. Therefore, in the design, the value of $K_{0,s}$ (K_0 without increase) should be used conservatively even when using the displacement method for encased column installation.

The area replacement ratio ($a_E = A_c/A_E$) is initially defined, which is the relationship between the column area A_c and its influence area A_E, i.e., the area of the unit cell, as sche-matically shown in Figure 2.1 for a group of encased columns installed in a triangular pattern.

The area of influence A_E is calculated by means of Equation (2.1), and the value of d_e (diameter of the area of influence) is taken as 1.05s (for triangular mesh), 1.13s (for square mesh), and 1.29s (for hexagonal mesh), where "s" is the center-to-center spacing between the columns. In this way, the area of influence is given by:

$$A_E = \pi \frac{d_e^2}{4} \tag{2.1}$$

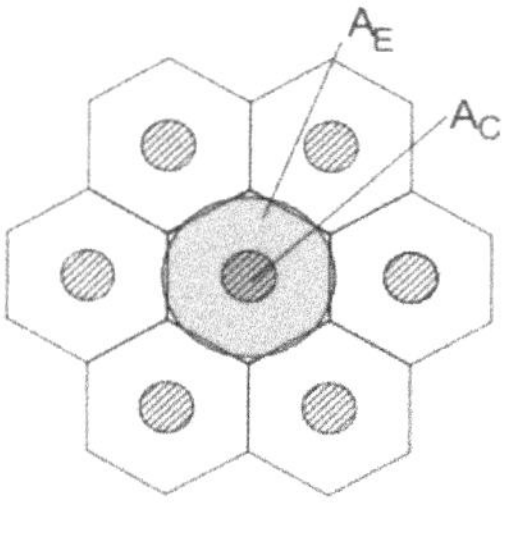

'unit cell'
$a = A_C / A_E$

Figure 2.1 Unit cell idealization: A_C = column area and A_E = unit cell area

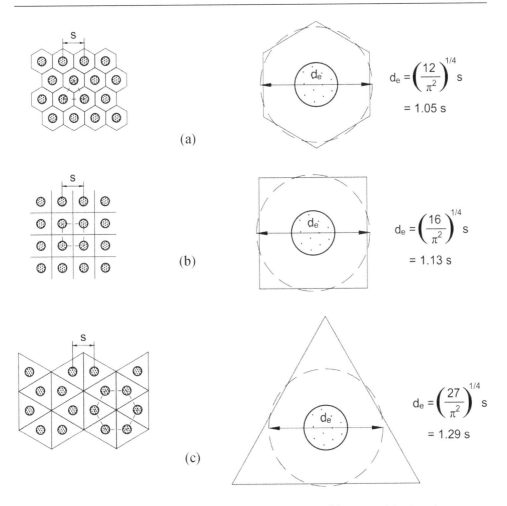

$$d_e = \left(\frac{12}{\pi^2}\right)^{1/4} s$$

$$= 1.05\, s$$

$$d_e = \left(\frac{16}{\pi^2}\right)^{1/4} s$$

$$= 1.13\, s$$

$$d_e = \left(\frac{27}{\pi^2}\right)^{1/4} s$$

$$= 1.29\, s$$

Figure 2.2 Common columns installation pattern: (a) hexagonal; (b) square; (c) triangular patterns (adopted from Balaam and Booker, 1981)

The column cross-sectional area A_c is simply calculated using the column diameter, d_c, which is typically equal to 0.80 m, but may vary as required. Figure 2.2 illustrates the triangular, square, and hexagonal meshes used in design. The square mesh pattern is the most commonly used, followed by the triangular one.

2.3 Vertical design: "analytical calculation"

The analytical model developed by Raithel (1999) and Raithel and Kempfert (2000) uses the unit cell concept, thus addressing the problem in the condition of axial symmetry with boundary conditions illustrated in Figure 2.3. A singular encased column and its surrounding soil are analyzed and the vertical stress equilibrium is developed, thus allowing to calculate the vertical and horizontal deformations separately on the encased column and surrounding soil.

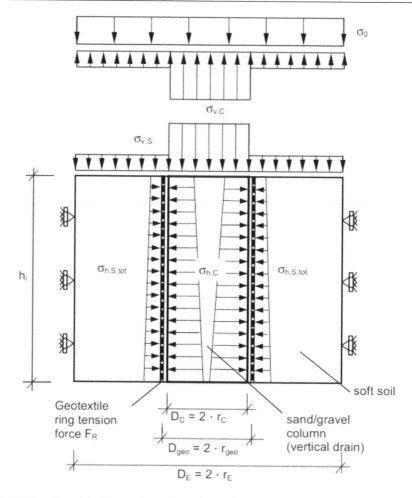

Figure 2.3 Unit cell model of encased granular column after Raithel and Kempfert (2000)

2.3.1 Equilibrium: "stresses and forces involved"

The total stress equilibrium between the embankment load $\Delta\sigma_0$ and the vertical stresses acting on the column $\Delta\sigma_{v,c}$ and the soil $\Delta\sigma_{v,s}$ is described by Equation (2.2) as follows:

$$\Delta\sigma_0.A_E = \Delta\sigma_{v,c}.A_c + \Delta\sigma_{v,s}.(A_E - A_c) \qquad (2.2)$$

where:

$\Delta\sigma_0$ = embankment total vertical stress, equal to the embankment-specific weight times the height of embankment;

$\Delta\sigma_{v,c}$ = vertical stress on top of the column;

$\Delta\sigma_{v,s}$ = vertical stress on top of the surrounding soil.

The vertical stress due to embankment loading ($\Delta\sigma_0$) and the unit weight of the surrounding soil and column material produces an increase in the horizontal (radial) stresses in the column and the surrounding soil, called $\Delta s_{h,c}$ and $\Delta\sigma_{h,s}$, respectively. The values of $\sigma_{v,0,c}$ and $s_{v,0,s}$ are the initial *in situ* vertical stresses in the column and the soil, respectively. These variables are presented in Equations (2.3) and (2.4). The value of $K^*_{0,s}$ shown in in Equation (2.4) must be replaced by $K_{0,s}$ if the excavation method is used.

$$\Delta\sigma_{h,c} = \Delta\sigma_{v,c}.K_{a,c} + \Delta\sigma_{v,0,c}.K_{a,c} \qquad (2.3)$$

$$\Delta\sigma_{h,s} = \Delta\sigma_{v,s}.K_{0,s} + \Delta\sigma_{v,0,s}.K^*_{0,s} \qquad (2.4)$$

The hoop ("ring") force developed on the geosynthetic encasement is given by F, which is calculated based on the hoop strain and the tensile stiffness modulus J of the geosynthetic encasement, provided by the manufacturer (Equation (2.5)). Note that J is time-dependent due to encasement creep and decreases with time (i.e., $J = J(t)$). The hoop strain can be calculated from the variation of the geosynthetic encasement radius (Δr_{geo}) and the original encasement radius (r_{geo}), as shown in Equation (2.5). The circumferential force given by Equation (2.5) can be transformed into horizontal (radial) stress acting on the geosynthetic encasement $\Delta\sigma_{h,geo}$, as shown in Equation (2.6).

$$F = J.\frac{\Delta r_{geo}}{r_{geo}} \qquad (2.5)$$

$$\Delta\sigma_{h,geo} = \frac{F}{r_{geo}} \qquad (2.6)$$

Equation (2.7) defines the value of $\Delta\sigma_{h,diff}$, which is the net horizontal stress resulting from the horizontal stress acting on the column ($\Delta\sigma_{h,c}$), the horizontal stress on surrounding soil ($\Delta\sigma_{h,s}$), and the horizontal stress developed on the geosynthetic encasement ($\Delta\sigma_{h,geo}$).

$$\Delta\sigma_{h,diff} = \Delta\sigma_{h,c} - (\Delta\sigma_{h,s} + \Delta\sigma_{h,geo}) \qquad (2.7)$$

This stress difference $D\sigma_{h,diff}$ causes the column to expand. The radial horizontal deformation Δr_c and the settlement on soil S_s can be calculated according to Equations (2.8) and (2.9), respectively, noting that the method considers the column settlement (S_c) as equal as the settlement on surrounding soil (S_s). Equation (2.8) is resulting from the approach of Ghionna and Jamiolkowski (1981) for calculation of the horizontal deformation of a hollow cylinder subjected to vertical loading.

$$\Delta r_c = \frac{\Delta\sigma_{h,diff}}{E^*}.\left(\frac{1}{a_E} - 1\right).r_c \qquad (2.8)$$

where:

$$E^* = \left(\frac{1}{1-\nu_s} + \frac{1}{1+\nu_s}.\frac{1}{a_E}\right).\frac{(1+\nu_s).(1-2\nu_s)}{(1-\nu_s)}.E_{oed,s}$$

v_s = Poisson's ratio of soft soil; and

$E_{oed,s}$ = constrained modulus of soft soil.

$$S_s = \left(\frac{\Delta\sigma_{v,s}}{E_{oed,s}} - 2.\frac{1}{E^*}.\frac{v_s}{1-v_s}.\Delta\sigma_{h,diff} \right).h_0 \tag{2.9}$$

h_0 = initial column's length.

$$\Delta r_c = \Delta r_{geo} + (r_{geo} - r_c) \tag{2.10}$$

r_0 = initial radius of the column.

Equation (2.10) is obtained from a purely geometric correlation. Thus, the incremental calculation of the settlement must be done by updating the values of h_0 and r_0. The variation of the radius of the column, Δr_c, is given by Equation (2.11) as follows:

$$E_{oed,s} = E_{oed,ref}.\left(\frac{p^* + c'.\cot\phi}{p_{ref}} \right)^m \tag{2.11}$$

Based on the intrinsic hypothesis of the method, the settlement at the top of the column is equal to the settlement in the surrounding soil, Equation (2.12):

$$S_c = S_s \tag{2.12}$$

According to the assumption given by Equation (2.12), Equation (2.13) is obtained with the variable Δr_c given by Equation (2.14).

$$\left[\frac{\Delta\sigma_{v,s}}{E_{oed,s}} - \frac{2}{E^*}.\frac{v_s}{1-v_s} \left[\begin{array}{l} K_{a,c}.\left(\frac{\Delta\sigma_v}{a_E} - \frac{1-a_E}{a_E}.\Delta\sigma_{v,s} + \Delta\sigma_{v,0,c} \right) \\ -K_{0,s}^*.\Delta\sigma_{v,s} - K_{0,s}.\sigma_{v,0,s} + \frac{J.(r_{geo} - r_c)}{r_{geo}^2} - \frac{J.\Delta r_c}{r_{geo}^2} \end{array} \right] \right].h = \left[1 - \frac{r_c^2}{(r_c + \Delta r_c)^2} \right].h \tag{2.13}$$

$$\Delta r_c = \frac{K_{a,c}.\left(\frac{\Delta\sigma_v}{a_E} - \frac{\Delta\sigma_{v,s}.(1-a_E)}{a_E} + \Delta\sigma_{v,0,c} \right) - K_{0,s}.\Delta\sigma_{v,s} - K_{0,s}^*.\Delta\sigma_{v,0,s} + \frac{J.(r_{geo} - r_c)}{r_{geo}^2}}{\frac{E^*}{\left(\frac{1}{a_E} - 1 \right).r_c} + \frac{J}{r_{geo}^2}} \tag{2.14}$$

where:

$K_{a,c}$ = active earth pressure of column material.

Substituting Equation (2.14) in Equation (2.13), only the value of $\Delta\sigma_{v,s}$ is indeterminate; Equation (2.13) must then be solved by an iterative process. In Equations (2.13) and (2.14), the value of J is the (time-dependent) "ring" tensile stiffness modulus of the geosynthetic encasement. For design purposes, $r_c = r_{geo}$ can be used in Equations (2.11), (2.13), and (2.14), assuming the encasement's radius is equal to the column's radius.

2.3.2 Soil constrained modulus $E_{oed,s}$

The value of the soil constrained modulus ($E_{oed,s}$) can be determined as a function of the vertical effective stress level in the middle of the soil layer, p*, as shown in Equation (2.15).

$$E_{oed,s} = E_{oed,ref} \cdot (p^* / p_{ref})^m \qquad (2.15)$$

where:
$E_{oed,ref}$ = reference constrained modulus of soil (see Fig. 2.4);
p* = effective vertical stress in middle of soft soil layer (see Equations (2.16) and (2.17));
p_{ref} = reference effective vertical stress (see Fig. 2.4);
m = exponent coefficient.

The value of p* to be considered can be a function of the vertical effective stress in the soil before loading (p_1^*) and vertical effective stress in the soil after loading (p_2^*), that is, before applying $\Delta\sigma_0$ (p_1^*) and after applying Δs_0 (p_2^*). The value of p* can be obtained by either Equation (2.16) or Equation (2.17). Equation (2.16) results in the net loading applied, and its use is preferred.

$$p^* = \frac{p_2^* - p_1^*}{\ln\left(\dfrac{p_2^*}{p_1^*}\right)} \qquad (2.16)$$

$$p^* = \frac{p_2^* + p_1^*}{2} \qquad (2.17)$$

The value of the constrained soil modulus ($E_{oed,s}$) can also be obtained by means of Equation (2.18) in order to consider the effective cohesion of the soil c'.

$$E_{oed,s} = E_{oed,ref} \cdot \left(\frac{p^* + c'.\cot\phi}{p_{ref}}\right)^m \qquad (2.18)$$

The effective cohesion in very soft soils is usually quite low, thus c' = 0 is a common design hypothesis. For the example presented in Figure 2.4, it is seen that $E_{oed,s} \approx 600$ kPa, $p_{ref} \approx 100$ kPa, and m ≈ 0.86. It should be noted that Raithel (1999) and Raithel and Kempfert (1999) present a procedure to consider multiple soil layers.

Although Raithel and Kempfert's (2000) method considers an equal settlement on column and soil, these settlements have shown to slightly differ from the numerical analyses and field measurements (Alexiew et al., 2012; Raithel et al., 2012). The results obtained by the analytical method, however, provided a good general agreement with results of numerical analysis and field data (e.g., Almeida et al., 2013; Riccio et al., 2012).

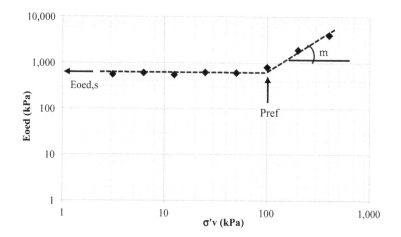

Figure 2.4 Variation of constrained modulus with vertical stress – values of $E_{oed,s}$, P_{ref}, and m

2.3.3 Geosynthetic tensile force

The geosynthetic tensile force obtained by Equation (2.5) must be compared with the allowable tensile force of the geosynthetic material (F_{all}) that is the maximum tension that the geosynthetics withstands in the long term. Reduction and safety factors have to be applied as presented in Equation (2.19) to evaluate F_{all} from F_0.

$$F_{all} = \frac{F_0}{RF_f . RF_{dm} . RF_{amb} . RF_{joint} . FOS} \tag{2.19}$$

where:

F_0 = tensile resistance in quick wide strip test (nominal resistance from catalog, it is often also called UTS – Ultimate Tensile Strength);
RF_f = reduction factor for creep;
RF_{dm} = reduction factor for mechanical damage;
RF_{amb} = reduction factor for chemical and environmental damages;
RF_{joint} = reduction factor for joints/seams, if exists;
FOS = factor of safety accounting for uncertainties in production and data extrapolation.

Information regarding reduction factors and typical values can be found, e.g., in German Recommendations EBGEO (2011) and British Standard BS 8006 (2010). Note that they can differ for various polymers and types of encasements; thus, precise values have to be provided by the geosynthetic manufacturer. The same is valid for the time-dependent ring tensile stiffness J.

2.4 Vertical design: consolidation analysis

In section 2.3, an analytical calculation was presented allowing computing the maximum settlement ($S_c = S_s$) of an embankment over a GEC system. However, in most cases it is important to determine the settlement development over time during both the embankment construction and the post-construction periods due to consolidation. Han and Ye (2002) developed an analytical method to calculate the degree of consolidation, originally proposed for the ordinary stone columns (Barron, 1948), but it could be applied to the encased columns, since the geosynthetic encasement allows for water flow.

Han and Ye (2002) assume the following hypotheses to develop their formulation:

- There is no vertical flow of water in the surrounding soil, either in the smeared zone or in the intact zone;
- Each column has a circular area of influence, and the granular columns are completely saturated, as is the surrounding soil;
- The granular column and the surrounding soil deform vertically, and the vertical deformations of both are equal, occurring in any depth;
- The coefficients of compressibility of the smeared and non-smeared zones are equal;
- The loading is applied instantly and considered constant during the consolidation period;
- The vertical stresses inside the columns and in the surrounding soil are considered uniform and constant along the depth;
- The excess pore pressure inside the column is uniform and constant along the radius column;

The settlement over time, S(t), can be calculated by Equation (2.20) through:

$$S(t) = S_c.U(t) \qquad (2.20)$$

where:

S_c = settlement on column or on surrounding soil ($S_s = S_c$) calculated as shown in section 2.3 considering the time-dependent tensile modulus $J = J(t)$ for a given time;

$U(t) = U$ = degree of consolidation due to radial water flow at a given time calculated using Han and Ye's (2002) method as follows:

$$U = 1 - e^{\frac{-8}{F_m'}.T_{rm}} \qquad (2.21)$$

where:

F_m' = consolidation function (see Equation (2.24));
T_{rm} = modified time factor.

The value of the modified time factor (T_{rm}) is determined by Equation (2.22) and is dependent on the modified radial consolidation coefficient (c_{rm}), given by Equation (2.23).

$$T_{rm} = \frac{c_{rm}.t}{d_e^2} \qquad (2.22)$$

where:

d_e = diameter of the unit cell (see Fig. 2.2);
t = time elapsed after application of the load; and

$$c_{rm} = \frac{k_r}{\gamma_w}.\frac{m_{v,c}(1-a_E)+m_{v,s}.a_E}{m_{v,s}.m_{v,c}.(1-a_E)} = c_r.\left(1+n_s.\frac{1}{N^2-1}\right) \qquad (2.23)$$

Where:

k_r = soil horizontal (or radial) permeability;

γ_w = specific unit weight of water;

$m_{v,c}$ = volumetric compressibility coefficient of column;

$m_{v,s}$ = volumetric compressibility coefficient of surrounding soil;

a_E = area replacement ratio as defined earlier;

c_r = coefficient of horizontal consolidation (due to radial flow);

n_s = stress concentration factor – relation of the vertical stress on top of the column and vertical stress on the top of the surrounding soil at the end of primary consolidation. According to the Han and Ye (2002), n_s can also be expressed as the ratio between the volumetric coefficients of soil and column, i.e., $n_s = m_{v,s}/m_{v,c}$. The recommended value for n_s, considering encasement effect, in the relation given by $\Delta\sigma_{v,c}/\Delta\sigma_{v,s}$, can be precisely estimated by EBGEO (2011).

N = ratio between the diameter of the unit cell and the diameter of the column, N = d_e/d_c.

The variable F'_m appearing in Equation (2.21) is determined as follows:

$$F'_m = \frac{N^2}{N^2-1} \cdot \left(\ln\frac{N}{S} + \frac{k_r}{k_s} \cdot \ln S_{mr} - \frac{3}{4} \right)$$
$$+ \frac{S_{mr}^2}{N^2-1} \cdot \left(1 - \frac{k_r}{k_s}\right) \cdot \left(1 - \frac{S_{mr}^2}{4N^2}\right) + \frac{k_r}{k_s} \cdot \frac{1}{N^2-1} \cdot \left(1 - \frac{1}{4N^2}\right) + \frac{32}{\pi^2}\left(\frac{k_r}{k_c}\right)\cdot\left(\frac{H}{d_c}\right)^2 \tag{2.24}$$

where:

k_s = permeability coefficient of soil in the smeared zone;

k_c = permeability coefficient of column material;

H = longest drainage path due vertical flow;

$S_{mr} = d_s/d_c$ = ratio between the smeared zone diameter (d_s) and the column's diameter (d_c).

As shown above, the formulation presented by Han and Ye (2002) does not consider the additional contribution of the vertical drainage, which is relatively small, as the water flow in the vertical direction is much smaller than in the radial direction.

2.5 Vertical design: column layout

The effect of the column layout is analyzed in this section. A comparison between the triangular ($d_e = 1.05s$), square ($d_e = 1.13s$), and hexagonal ($d_e = 1.29s$) meshes is presented, where d_e is the diameter of influence area and s the spacing between axes of columns, as shown in Figure 2.2. The area replacement ratio a_E will affect significantly the foundation settlement and the hoop tensile force developed in the geosynthetic encasement, and the cost of the foundation is directly related to the area ratio a_E. The greater the a_E the higher the cost, as more columns need to be implemented.

Column spacing, s, can be defined from the desired area ratio, column diameter, and type of mesh to be adopted (triangular, square, or hexagonal). In Equation (2.25), the area of influence a_E is presented as a function of the column area A_c and the area replacement ratio a_E.

$$A_E = \frac{A_c}{a_E} \qquad (2.25)$$

This results in Equation (2.26) as follows:

$$d_e^2 = \frac{d_c^2}{a_E} \qquad (2.26)$$

In this way, Equation (2.26) results in Equation (2.27):

$$s_m^2.s^2 = \frac{d_c^2}{a_E} \qquad (2.27)$$

The value of s is given by:

$$s = \frac{d_c}{s_m}.\sqrt{\frac{1}{a_E}} \qquad (2.28)$$

Figure 2.5 shows the required column spacing to produce the area ratios equal to 0.10, 0.20, and 0.30 considering different mesh layouts. The results are presented for the column diameter (d_c) equal to 0.80 m. It is observed that the triangular mesh is the most efficient pattern because it is possible to achieve the desirable area ratio (a_E) with larger spacing compared with square and hexagonal layouts.

Figure 2.6 presents results of settlements as function of different values of area ratio (a_E) and using parameters presented in Tables 2.1 and 2.2. In this figure, the soft soil thickness (H_s) is 8.0 m, the tensile stiffness modulus (J) is 2000 kN m^{-1}, and the imposed load ($\Delta\sigma_0$) is 100 kPa. It is observed that the increase of a_E results in significant reduction of settlement and its influence is more pronounced when a_E value varies between 0.10 and 0.20.

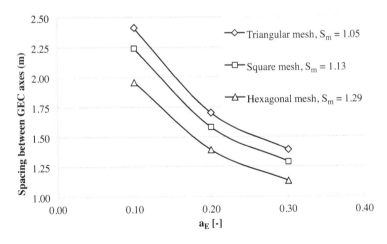

Figure 2.5 Spacing between columns axis (s) depending on mesh type (s_m), area ratio (a_E), and column diameter (d_c)

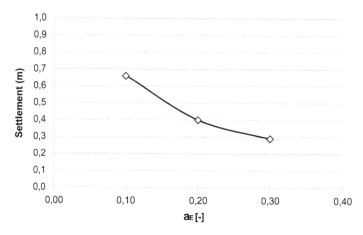

Figure 2.6 Settlements as function of area ratio a_E, where: H_s = 8.0 m; J = 2000 kN m^{-1}; $\Delta\sigma_0$ = 100 kPa; other parameters as presented in Tables 2.6 and 2.7

Table 2.1 Parameters of soft soil and column fill (Riccio et al., 2016)

Parameter	Value
c_s' (soft soil effective cohesion)	2.0 kPa
ϕ_s' (soft soil effective friction angle)	28°
$E_{oed,s,ref}$ (reference constrained modulus of soft soil)	1000 kPa
v_s (Poisson's ratio of soft soil, drained condition)	0.30
m (exponent, variation of $E_{oed,s}$ with vertical effective stress)	0.84
P_{ref} (reference vertical effective stress)	100 kPa
γ_s' (soft soil submerged unit weight)	4 kN m^{-3}
ϕ_c' (effective friction angle of column filling material)	30°
γ_c' (column fill material submerged unit weight)	8 kN m^{-3}

Table 2.2 Geometrical parameters (Riccio et al., 2016)

Parameter	Value
a_E (area ratio)	0.10; 0.20; 0.30
H_s (soft soil thickness)	4 m; 8 m; 16 m
J (geosynthetic stiffness modulus)	2000; 4000; 6000 kN m^{-1}
r_{geo} (geosynthetic encasement radius)	0.40 m
r_c (column radius)	0.40 m
s (columns center to center spacing)	Variable

Riccio et al. (2016) analyzed a hypothetical case of an embankment over encased columns with parameters presented in Tables 2.1 and 2.2. The soil parameters used are typical of the west zone of the city of Rio de Janeiro, and the imposed loads (embankment height multiplied by the natural specific weight of the embankment) are variables. The parameters of the column fill material are also typical for encased granular column soil. The "ring" stiffness modulus J of geosynthetic encasement was changed allowing a parametric analysis. The calculations were performed by means of Raithel and Kempfert's (2000) method described in section 2.3.

Figure 2.7 shows the variation of vertical stress on surrounding soil ($\Delta\sigma_{v,s}$) as a function of soft soil thickness (H_s) and applied load ($\Delta\sigma_0$) using parameters presented in Tables 2.1 and 2.2. The area ratio, a_E, is 0.30 and the stiffness modulus of geosynthetic encasement, J, is 2000 kN m^{-1}. As seen in Figure 2.7, the higher the soft soil thickness, the greater is the vertical stress acting on surrounding soil. Considering an applied load of 150 kPa, the vertical stress acting on soil is around 15 kPa for a soft soil thickness of 4.0 m; on the other hand, for a soft soil thickness equal to 16.0 m, the vertical stress on soft soil is around 25 kPa. Therefore, the GEC column has the effect of reducing the vertical stress on soil. This effect will be more pronounced by increasing the stiffness modulus (J) (Riccio et al., 2016).

In Figure 2.8, variation of soil vertical stress $\Delta\sigma_{v,s}$ is plotted as a function of a_E and J, while soil thickness and imposed load are kept constant at 8.0 m and 100 kPa, respectively.

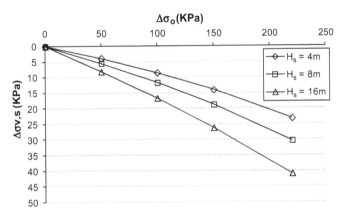

Figure 2.7 Vertical stress on soil ($\Delta\sigma_{v,s}$) as a function of loading ($a_E = 0.30$, J = 2000 kN m^{-1}), Riccio et al. (2016)

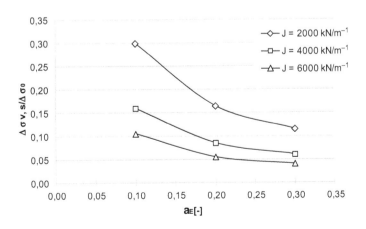

Figure 2.8 Variation of $\Delta\sigma_{v,s}$ as a function of a_E and J, soil thickness equal to 8.0 m and $\Delta\sigma_0$ = 100 kPa

As seen, both a higher area ratio a_E and modulus J reduce the total soil vertical stress, which is plausible from the engineering point of view and corresponds to some simplified analyses in Alexiew *et al.* (2005). Note that the benefit of increasing a_E in the range of 10–20% is more significant than for $a_E > 20\%$. This corresponds to the present experience from executed projects (Alexiew *et al.*, 2012; Alexiew and Thomson, 2013; Alexiew *et al.*, 2016) and confirms recommendations in EBGEO (2011).

It should be noted that the choice of both a_E and modulus J to control settlement and stability is a matter of optimization, as similar results of final settlements, for instance, can be achieved using different column layouts. Technical, economic, and environmental factors have to be taken into account inclusive of, e.g., total costs, number of installation rigs, construction time, and carbon footprint. More comments, details, and optimization recommendations can be found in Alexiew and Thomson (2014).

2.6 Alternative calculation methods

Alternative methods for foundation designs of GEC systems were proposed by Pulko *et al.* (2011); Zhang *et al.* (2011); Castro and Sagaseta (2013); and Zhang and Zhao (2015). These methods are compared with Raithel and Kempfert's (2000) method, as can be seen in Table 2.3.

2.7 Global stability: plane strain analysis

An important aspect of an embankment construction over soft soil treated with encased granular columns is to verify the stability of embankment by calculating the factor of safety. This section presents a method for a 2D global stability analysis using either the finite element simulation or the limit equilibrium method. The numerical analysis here is performed based on the methodology proposed by Raithel and Henne (2000) and Tan *et al.* (2008) and successfully used by Hosseinpour *et al.* (2017).

Tan *et al.* (2008) proposed a simplified methodology for 2D plane strain analysis of the conventional granular columns, in which the unit cell of granular column is converted into a wall to obtain the equivalent plane strain column width, as illustrated in Figure 2.9.

According to Tan *et al.* (2008), the following relationship is used to calculate the half of the column width b_c in 2D plane strain simulation:

$$b_c = B \frac{r_c^2}{R^2} \qquad (2.29)$$

where:
r_c = radius of the column (see Fig. 2.9a);
b_c = half of the width of the equivalent wall (see Fig. 2.9b);
R = radius of the influence area (see Fig. 2.9a);
B = half of the equivalent width in 2D plane strain condition (see Fig. 2.9b).

Equation (2.30) gives the relation between R and B, as follows:

$$R = 1.13B \qquad (2.30)$$

This approach, however, is not calibrated for the influence of the confining support provided by the geosynthetic encasement, thus the increase of column strength is not taken into account.

Table 2.3 Characteristics and hypothesis of GEC calculation methods

Material behavior and parameter	Analytical Method				
	Raithel and Kempfert (2000)	Pulko et al. (2011)	Zhang et al. (2011)	Castro and Sagaseta (2013)	Zhang and Zhao (2015)
Column	Elastic-plastic (ϕ, ψ, a_E, K_a)	Elastic-plastic (E, ϕ, ψ, v)	Elastic-plastic (E, ϕ, ψ, v, a_E)	Elastic-plastic (E, ϕ, ψ, v, a_E)	Elastic-plastic (E, ϕ, ψ, v, a_E, K_0)
Soft soil	Drained-Elastic (E_{oed}, ϕ, c, v, K_0)	Drained-Elastic (E, ϕ, c, v)	Drained-Elastic (E_{oed}, ϕ, c, v)	Undrained-Elastic (E, ϕ, c, v)	Drained-Elastic (E, ϕ, c, v)
Encasement	Elastic (J, r)	Elastic (J)	Elastic (J)	Elastic-plastic (J, T_{max})	Elastic (J, t)
Limitations of each method	Geosynthetic plasticity Soil inhomogeneity Drained analysis Constant geosynthetic tensile stress	Geosynthetic plasticity Soil inhomogeneity Drained analysis Column installation	Geosynthetic plasticity Soil inhomogeneity Drained analysis Column installation	Soil inhomogeneity Column installation	Geosynthetic plasticity Soil inhomogeneity Drained analysis Column installation

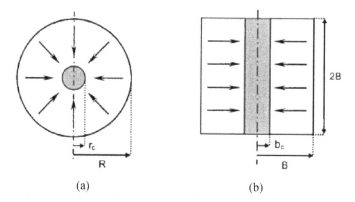

(a)   (b)

Figure 2.9 Tan et al. (2008) proposal for geometric transformation of 3D to 2D condition: (a) original condition (axisymmetric); (b) transformed condition (plane strain)

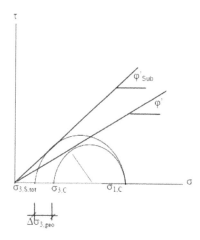

Figure 2.10 Horizontal stresses on the border of geosynthetic encased column (adapted from Raithel and Henne, 2000)

In order to consider the contribution of the geosynthetic encasement in the shear strength of the columns, the proposal of Raithel and Henne (2000), recommended in EBGEO (2011), may be used. In this proposal, the friction angle of the column filling material (ϕ') is replaced by an increased friction angle (ϕ'_{sub}), which takes into account the confining effect of the geosynthetic encasement. Figure 2.10 illustrates the Mohr circles obtained from the horizontal stresses located at a point on the border of the encased column. A similar concept (Malarvizhi and Ilamparuthi, 2008) increases c' instead of ϕ'.

The values of $\sigma_{3,s,tot}$, $\sigma_{3,c}$ and $\Delta\sigma_{1,c}$ shown in Figure 2.10 are defined as follows:

- $\sigma_{3,s,tot}$ = total horizontal stress (or resulting stress) that acts in the soil due to compression of the column and relieved by the confinement produced by the encasement;
- $\Delta\sigma_{3,geo}$ = horizontal (radial) stress supported by geosynthetic;
- $\sigma_{3,c}$ = horizontal (radial) stress inside the column.

The development of the solution assumes that the equilibrium of horizontal stresses is given by Equation (2.31) below:

$$\sigma_{3,s,tot} + \Delta\sigma_{3,geo} = \sigma_{3,c} \tag{2.31}$$

The solution considers the relationship between the major stresses ($\sigma_{1,c}$, vertical stress) and minor ($\sigma_{3,c}$, horizontal stress) in the inner part of the column, Equation (2.32).

$$\sigma_{1,c} = \sigma_{3,c} \cdot \frac{1+\sin\phi'}{1-\sin\phi'} + 2c' \cdot \frac{\cos\phi'}{1-\sin\phi'} \tag{2.32}$$

Considering the null cohesion and the relation given by Equation (2.31) yields:

$$\sigma_{1,c} = \sigma_{3,c} \cdot \frac{1+\sin\phi'}{1-\sin\phi'} \tag{2.33}$$

$$\sigma_{1,c} = (\sigma_{3,s,tot} + \Delta\sigma_{3,geo}) \cdot \frac{1+\sin\phi'}{1-\sin\phi'} \tag{2.34}$$

By developing Equations (2.33) and (2.34), the equivalent (or substitute) effective friction angle is calculated, which considers the effect of confinement provided by geosynthetic encasement. The value of ϕ'_{sub} is given by Equation (2.35) as follows:

$$\sin\phi'_{sub} = \frac{\left(\dfrac{1+\sin\phi'}{1-\sin\phi'}\right) + \left(\dfrac{\Delta\sigma_{3,geo}}{\sigma_{3,c}}\right) - 1}{\left(\dfrac{1+\sin\phi'}{1-\sin\phi'}\right) - \left(\dfrac{\Delta\sigma_{3,geo}}{\sigma_{3,c}}\right) + 1} \tag{2.35}$$

The stresses $\Delta\sigma_{3,geo}$ and $\sigma_{3,c}$ can be obtained by Equation (2.6) ($\Delta\sigma_{h,geo}$) and Equation (2.3) ($\sigma_{h,c}$), respectively. Another way to obtain these stresses is by means of numerical analysis, in which the unit cell approach and axisymmetric type analysis is valid.

One alternative, instead of using substitute higher c' or ϕ', is to perform stability analyses using the higher pressure on top of GECs calculated from the vertical design, which then "automatically" increases the shear strength in the columns (Raithel et al., 2012). However, this option is more time consuming than calculations using the methods proposed by Raithel and Henne (2000) or Malarvizhi and Ilamparuthi (2008).

The ϕ'_{sub} (or c'_{sub}) are the increased shear parameters to be considered for the granular filling material of the encased columns in 2D plane strain analysis. Therefore, the following steps are suggested for global stability purposes (ULS via limit equilibrium methods or numerical analysis) and, if needed, for a more precise calculation of global deformations (numerical analysis, depending on the model):

- Perform the 3D geometry transformation to 2D condition by means of Tan et al.'s (2008) method proposed for ordinary column;
- Calculate the increased shear parameters ϕ'_{sub} or c'_{sub}. If necessary, the unit cell analysis should be performed to determine the required parameters;

- Replace the original ϕ' or c' of the column fill (usually $c' = 0$) by ϕ'_{sub} as per Raithel and Henne's (2000) or c'_{sub} as per Malarvizhi and Ilamparuthi's (2008) proposal;
- If necessary, divide the length of the column into sub-layers by obtaining more than one value for ϕ'_{sub} or c'_{sub} along the column's length. This procedure should be adopted when there is a significant variation of $\Delta\sigma_{3,geo}$ and $\sigma_{3,c}$ along the length of the column. This approach becomes feasible when using numerical analysis.

If the global stability calculated shows a low factor of safety, the best way to improve the stability is to add a basal geosynthetic reinforcement on the top of the GECs (EBGEO, 2011).

2.8 Final comments

2.8.1 Vertical and global designs

"Vertical" design (focusing on the Serviceability Limit State, SLS) and "global" design (focusing on the Ultimate Limit State, ULS) for embankments supported by GECs were discussed in this chapter. Vertical design allows for the calculations of settlements as well as tensile force in the geosynthetic encasement, and global design allows for global system stability and factor of safety calculations. Settlement calculations obtained from vertical design are referred to as basal settlements, assumed equal by the analytical method used herein, which in fact is not observed in practice (Almeida et al., 2014).

If the design is related to a fixed embankment top elevation, as in the case of roads and railways, then the embankment height needs to be increased to compensate for settlements. Therefore, vertical design calculations need to be performed again as embankment vertical stresses have increased, and this is in fact an interactive process (Almeida and Marques, 2013). Often some surplus of height at the beginning of the embankment construction is a proper solution Alexiew et al. (2012). However, in some geotechnical problems, such as stockyards, the height of embankment or applied vertical stress is the fixed value rather than the embankment top elevation.

If the water level is close to ground level, embankment submergence may occur after settlement stabilization, thus applied embankment vertical stresses need to be corrected, resulting also in an interactive process (Almeida and Marques, 2013).

2.8.2 Differential surface settlements

No mention was given yet to differential settlements on the top of the embankment (surface settlements) on GECs, which may be obtained, for instance, by numerical analyses. The critical height of embankment H_{crit}, often used in relation to piled embankments, may be defined as the height of embankment above which surface differential settlements do not take place.

Numerical calculations have shown that the analytical equations used to compute the critical height of embankment for piled embankments may be satisfactorily adopted also for GEC-supported embankments (Carreira et al., 2016). This equation, proposed by McGuire et al. (2012) and based on experimental and numerical studies, is given by:

$$H_{crit} = 1.15s' + 1.44d_c \tag{2.36}$$

where:

d_c = diameter of encased column as defined earlier; and s′ is given by Equation (2.37), for square mesh:

$$s' = \frac{\sqrt{(s_1^2 + s_2^2)}}{2} - \frac{d_c}{2}$$ (2.37)

where:

$s_1 = s_2$ are the spacing between the column axes.

Therefore, if a value of embankment height greater than H_{crit} is used, no differential surface settlements will take place. This check is on the safe side, because the "pointed support" provided by GECs is "softer" in comparison to rigid columns, thus resulting in smaller settlement differences at any level.

2.8.3 Proposed steps for design

In conclusion, the proposed steps for the design of a GEC-supported embankment are:

a. Adopt a value of GEC diameter d_c in accordance with the commercially available installation pipes; $d_c = 0.80$ m is probably the most widely used GEC diameter.

b. Adopt a distance "s" between columns GEC; it is suggested that the height of embankment H_{emb} is ideally greater than H_{crit} defined by Equation (2.36) so that surface differential settlements do not occur, if this is a requirement (not in the case of stockyards, for instance).

c. Perform vertical design, thus calculating the GEC tension force and settlements on both column and soft soil. Perform iterative calculations if a fixed top embankment elevation is required or if embankment submergence occurs.

d. Perform global design to ensure an adequate factor of safety, thus adequate global stability. Basal reinforcement may be added if increasing the factor of safety is required.

Chapter 3

Parameters used in pre-design charts and calculation examples

3.1 Introduction

This chapter presents the material properties used in the pre-design charts of the GECs-reinforced foundation inserted into Annexes I and II. Some calculation examples are also presented afterward to demonstrate how to use the pre-design charts. The charts provided are based on the analytical calculation method proposed by Raithel and Kempfert (2000) and explained in detailed in Chapter 2. The adopted soil parameters are deemed to be typical and cover a range of values normally found in projects involving soft soil treatment with GECs.

The main parameters affecting the results are the drained soft soil strength parameters (c' and ϕ'), the thickness of the soft soil layer (H_s), the total applied embankment load including the external live loads and/or external permanent loads ($\Delta\sigma_0$), the area replacement ratio (a_E), and the tensile stiffness modulus of the geosynthetic encasement in "ring" direction (J). In all cases shown herein, the radius of the geosynthetic encasement (r_{geo}) is 0.4 m, being equal to the column radius (r_c). The water table level was also assumed to be located on the top of the soft soil layer.

3.2 Limit state design

In order to perform a suitable analysis, two design approaches should be satisfied: the Ultimate Limit State (ULS, the system and its components are assumed in the stage of failure; strength-related parameters control the design) and the Serviceability Limit State (SLS, deformation-related parameters control the design). Different partial safety factors are applied in both cases depending on codes used in different countries. Roughly speaking for the ULS, they increase the actions and reduce the resistances. For the SLS, often all partial safety factors are set to one. A suggestion according to the factors is given in Table 3.1 for the purpose of orientation only.

The ULS approach considers partial safety factors to increase the unit weight of soil (f_s), external permanent load (f_t), and external live load (f_q). In the ULS approach, moreover, factor of safety is applied to reduce the tensile strength of the geosynthetic encasement (i.e., RF_f, RF_{dm}, RF_{amb}, RF_{joint}, and FOS) as described in section 2.3.3.

On the other hand, the SLS approach does not consider the increase in the soil unit weight ($f_s = 1.00$), external permanent load ($f_t = 1.00$), and external live load ($f_q = 1.00$) or the partial safety factor regarding the strength of the geosynthetic.

It is necessary to point out that the partial safety factors depend on the standard of each country; however, the factor of safety regarding the tensile strength of the geosynthetic reinforcement depends on the manufacturer.

Table 3.1 Partial factors of safety and reduction factors

Partial factor of safety or reduction factors		Ultimate Limit State (ULS)	Serviceability Limit State (SLS)
Load factors, increasing "action"	Unit weight of embankment fill	$f_s = 1.30$*	$f_s = 1.00$
	External permanent load	$f_t = 1.20$*	$f_t = 1.00$
	External live load	$f_q = 1.30$*	$f_q = 1.00$
Soil factors, decreasing "resistance"	Tangent of angle of internal friction	$f_{fi} = 1.20$	–
	Cohesion	$f_c = 1.5$	
Encasement factors, decreasing strength and/or tensile modulus	Applied in the encasement strength	Reducing strength, see section 2.3.3	Reducing tensile modulus
			–
Reduction factors	Accounting for:	Product-dependent RF_f	RF_{dm} in combination with isochronous curves
	Creep		
	Installation and mechanical damage	RF_{dm}	
	Environment	RF_{amb}	

* Suggested values (depends on the standard of each country).

It is important to highlight that for vertical design purposes, no partial factors of safety are applied, neither partial nor global. In case of factor of safety application (global design), this should be done depending on the country and the valid geotechnical codes.

3.3 Material parameters

The pre-design charts are organized in four groups: A, B, C, and D. Groups A and B are presented in Annex I and groups C and D are inserted in Annex II. Groups A and B consider a soft soil reference constrained modulus of 500 kPa, and groups C and D assume it equal to 1500 kPa.

Table 3.2 shows the other required material parameters used in the analyses. It is noted that the main differences between the groups are the values of soft soil constrained modulus (E_{oed}) and drained strength parameters (c' and ϕ'). Figure 3.1 illustrates the geometry of the hypothetical case considered for the analysis with material properties presented in Table 3.2.

3.4 Vertical design: charts utilization

For each soft soil thickness (i.e., $H_s = 5$, 10, 15, or 20 m), four groups of charts are presented in order to calculate the following values according to the total applied embankment stress ($\Delta\sigma_0$):

- Circumferential tensile force in the geosynthetic encasement (F);
- Normalized settlement with respect to the soft soil thickness (S/H_s);
- Vertical stress on soft soil ($\Delta\sigma_{v,s}$); and
- Vertical stress on top of the column ($\Delta\sigma_{v,c}$).

Note that some lines in the charts to obtain the above four variables (F, S/H_s, $\Delta\sigma_{v,s}$, $D\sigma_{v,c}$) end up earlier due to the 5% limitation of the maximum specific axial strain in the geosynthetic

Table 3.2 Parameters used in pre-design charts according to the analysis carried out

Properties	Group A (Annex I)	Group B (Annex I)	Group C (Annex II)	Group D (Annex II)
γ'_s (kN m⁻³)	4	4	4	4
γ'_c (kN m⁻³)	9	9	9	9
v_s (−)	0.4	0.4	0.4	0.4
m_s (−)	0.84	0.84	0.84	0.84
ϕ'_s (°)	25	28	25	28
ϕ'_c (°)	30	30	30	30
P_{ref} (kPa)	100	100	100	100
$K_{0,s} = 1 - \sin\phi'$	0.58	0.53	0.58	0.53
c'_s (kPa)	2	5	2	5
H_s (m)	5, 10, 15, 20	5, 10, 15, 20	5, 10, 15, 20	5, 10, 15, 20
$E_{oed, ref}$ (kPa)	500	500	1500	1500
a_E (%)	10, 15, 20	10, 15, 20	10, 15, 20	10, 15, 20
J (kN m⁻¹)	2000, 3500, 6500	2000, 3500, 6500	2000, 3500, 6500	2000, 3500, 6500

Subscripts "s" and "c" refer to the soft soil and column, respectively.

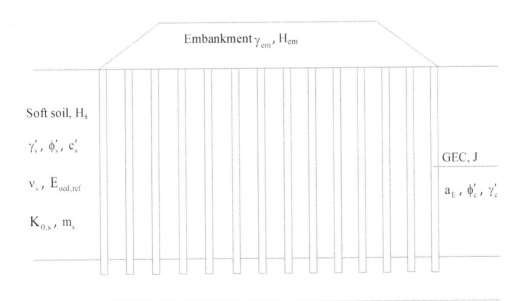

Figure 3.1 Hypothetical embankment on GECs-soft clay ground

encasement. If the selected value of $\Delta\sigma_0$ is not covered by any line, a greater tensile stiffness modulus of geosynthetic (J) or a higher area ratio necessarily need to be selected (a_E).

3.5 Calculation examples

Some examples are presented here aiming to calculate the desire variables mentioned above for a hypothetical GECs system. In practice, if the variables are not included in the charts, an interpolation can be applied whenever possible. This seems acceptable due to the pre-design character of the charts. For the final design, precise analytical and/or numerical calculations should be performed. The same is valid if the variables are quite different from those presented in the charts.

3.5.1 Example 1: embankment over GECs – "vertical design"

Consider an embankment constructed over a 10 m thick soft soil layer (H_s) improved by GECs as illustrated in Figure 3.2. It is necessary to perform a vertical design, thus determining the embankment settlement (leading design criteria to be met), the geosynthetic tensile force, the vertical stress on top of the column, as well as the vertical stress on the surrounding soft soil. The embankment is 4.0 m high (H_{em}) and the apparent unit weight of the compacted fill is equal to γ_{em} =19 kN m^{-3}. The geosynthetic encasement has a tensile stiffness modulus (J) of 2000 kN m^{-1} and the column diameter is 0.80 m. The column is assumed to be formed by sand material with effective angle of friction (ϕ'_c) equal to 30°. The reference constrained modulus ($E_{oed,ref}$), the effective cohesion (c'_s), and the drained angle of friction (ϕ'_s) of the soft soil are 500 kPa, 5 kPa, and 28°, respectively. The area replacement ratio (a_E) is also 10%.

Solution

The total embankment applied stress is:

$$\Delta\sigma_0 = 4m \times 19kN / m^3 = 76kPa$$

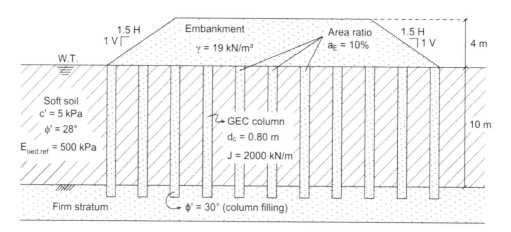

Figure 3.2 Schematic representation of example 1

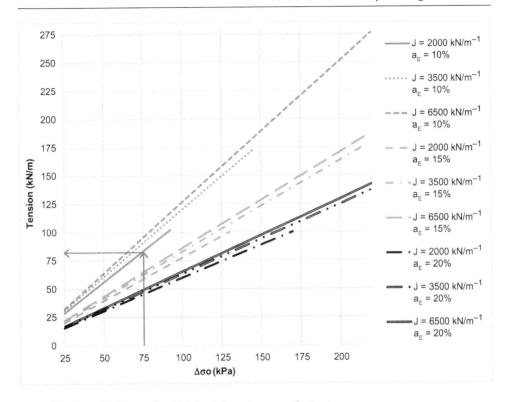

Figure 3.3 Chart 5B (Annex I) – Mobilized force in geosynthetic ring

Using Annex I and considering $E_{oed,ref}$ = 500 kPa, the pre-design charts shown in group B are used to determine the corresponding output variables. The charts shown in group B concern a soft soil layer with strength parameters c'_s = 5 kPa and ϕ'_s = 28°. Since the soft soil thickness is H_s = 10 m, the charts 5B, 6B, 7B, and 8B are used.

The circumferential tension (F) in the geosynthetic encasement is calculated by means of chart 5B (Annex I, group B) as illustrated in Figure 3.3.

Using chart 5B (i.e., Fig. 3.3) for $\Delta\sigma_0$ = 76 kPa and the curve corresponding to a_E = 10% and J = 2000 kN m^{-1}, the mobilized ring force in geosynthetic encasement is F = 85 kN m^{-1}. Chart 6B is utilized (Fig. 3.4) to calculate the settlement. Considering a_E = 10% and J = 2000 kN m^{-1}, the normalized settlement is S/H$_s$ = 0.075. Since the soft clay thickness is H$_s$ = 10 m, the calculated settlement value is 0.075 × 10 = 0.75 m. The soft soil vertical stress is obtained using chart 7B (Fig. 3.5) and the curve corresponding to a_E = 10% and J = 2000 kN m^{-1} results in a soil vertical stress of $\Delta\sigma_{v,s}$ = 15 kPa. The column vertical stress is determined from chart 8B (Fig. 3.6), and using the curve corresponding to a_E = 10% and J = 2000 kN m^{-1} results in a vertical stress on column of $\Delta s_{v,c}$ = 600 kPa. In the present example, the calculated relationship between $\Delta\sigma_{v,c}$ and $\Delta\sigma_{v,s}$ (i.e., known as stress concentration factor $n_s = \Delta\sigma_{v,c}/\Delta\sigma_{v,s}$) is quite high and equal to 40. However, field measurements indicate that this ratio is commonly much smaller, in the range typically between 3 and 8, mainly due to the presence of a working platform above the soft soil layer. Raithel and Kempfert's (2000) analytical solution, however, does not consider the existence of this granular platform layer, consequently theoretical n_s values are much higher.

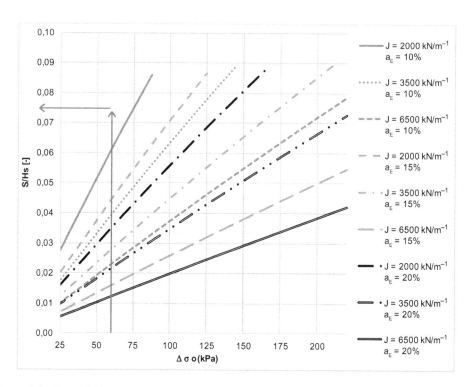

Figure 3.4 Chart 6B (Annex I) – Normalized settlement

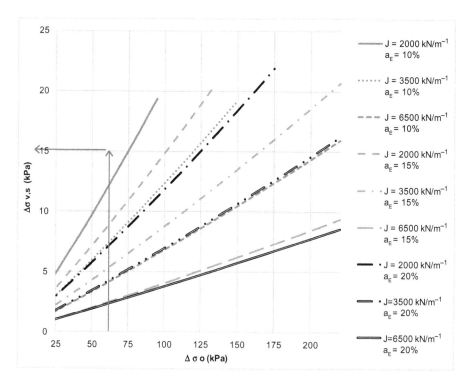

Figure 3.5 Chart 7B (Annex I) – Vertical stress on top of soft soil

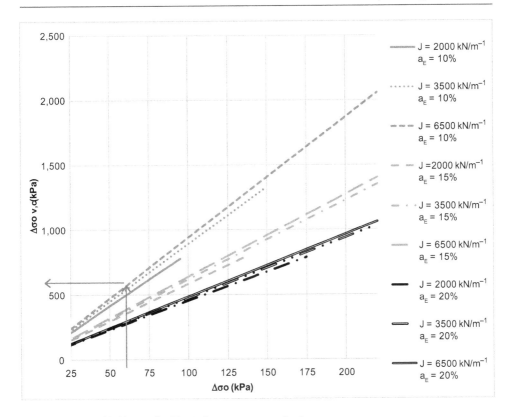

Figure 3.6 Chart 8B (Annex I) – Vertical stress on top of column

To perform the Serviceability Limit State and the Ultimate Limit State analyses, the partial factors of safety and reduction factors must be applied (see Table 3.1).

3.5.2 Example 2: stability analysis of embankment on GECs – "global design"

It is required to evaluate the short- and long-term global stability (global design) of a 4.0-m high embankment (H_{em}) over GECs-reinforced soft foundation with columns diameter equal to d_c = 0.8 m spaced s = 1.8 m in a square mesh, as shown in Figure 3.7. For this purpose, the method proposed by Raithel and Henne (2000) is used, in which the encased granular columns are converted into equivalent walls with an increased angle of internal friction to take into account the encasement confining effect.

The compacted embankment fill has an apparent unit weight equal to γ_{em} = 19 kN m^{-3} and drained angle of friction of ϕ' = 30°. The soft soil layer is 15 m thick, with the reference constrained modulus of $E_{oed,ref}$ = 1500 kPa, effective cohesion of c'_s = 2 kPa and drained angle of friction of ϕ'_s = 25°. The undrained shear strength of soft soil is S_u = 14 kPa on top with an increment of 1.0 kPa m^{-1} along the depth (S_u = 14 + 1.0z). The column filling material is sand with drained angle of friction equal to ϕ'_c = 30° and the apparent unit weight of γ_c = 19 kN m^{-3}. The geosynthetic encasement has a tensile stiffness modulus

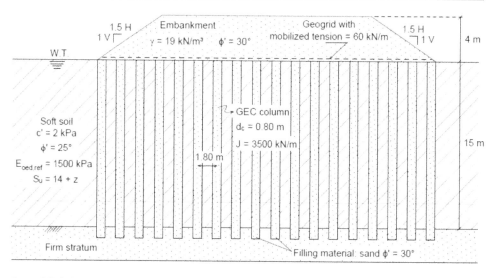

Figure 3.7 Schematic representation of example 2

(J) equal to 3500 kN m⁻¹. A basal geogrid is placed over the columns with 60 kN m⁻¹ effective (allowed) tensile force. The pre-design charts here are used aiming to calculate the following variables:

a.   Mobilized ring force in geosynthetic encasement;
b.   Settlement of the embankment;
c.   Vertical stress on column; and
d.   Vertical stress on surrounding soil.

Solution

The total applied stress imposed by the embankment is:

$$\Delta\sigma_0 = 4m \times 19kN / m^3 = 76kPa$$

From Annex II and considering soft soil constrained modulus equal to 1500 kPa, group C charts, developed for soil strength properties of $c_s' = 2$ kPa and $\phi_s' = 25°$, should be used. Among those, charts 9C, 10C, 11C, and 12C are selected since the soft soil thickness is $H_s = 15$ m. Considering 0.8 m diameter columns spaced s = 1.80 m in a square mesh, the area replacement ratio (a_E) is calculated as follows:

$d_e = 1.13s = 1.13 \times 1.80 = 2.03$ m
$A_E = 3.25$ m² (see Equation (2.1))
$d_c = 0.80$ m
$A_c = 0.50$ m²
$a_E = A_c/A_E = 0.15$ ($a_E = 15\%$).

The mobilized "ring" tensile force in the encasement is determined using chart 9C (Annex II). For the present case, considering $J = 3500$ kN m^{-1} and $a_E = 15\%$, the tensile force is $F = 50$ kN m^{-1}. Chart 10C (Annex II) gives a normalized settlement $S/H_s = 0.026$. Considering soft soil thickness of $H_s = 15$ m, the total embankment settlement is $S = 0.39$ m.

The vertical stress on soil is calculated using chart 11C, thus considering $J = 3500$ kN m^{-1}, $a_E = 15\%$, and $\Delta\sigma_0 = 76$ kPa, the vertical stress on the top of the soft soil is $\Delta\sigma_{v,s} = 20$ kPa.

Similarly, for $J = 3500$ kN m^{-1}, $a_E = 15\%$, and $\Delta\sigma_0 = 76$ kPa and using chart 12C, the vertical stress on the top of the encased column is $\Delta\sigma_{v,c} = 400$ kPa.

For global stability purposes, the column geometry needs to be transformed from 3D to 2D condition, which can be done using Tan et al.'s (2008) proposal, described in Chapter 2. Applying this to the geometrical conditions of the present example results in the equivalent column width $(2b_c)$ equal to 0.28 m as follows:

$r_c = 0.40$ m;
$d_e = 2.03$ m; $R = d_e/2 = 1.015$ m
$R = 1.13B$; $B = 1.015/1.13 = 0.90$ m
$b_c = 0.90 \ (0.40^2/1.015^2) = 0.14$ m (see Equation (2.29))

The spacing between the axis of the equivalent walls is $2B$, that is exactly equal to the center-to-center spacing between the columns $s = 1.80$ m. The clear distance between two adjacent equivalent walls $(2B-2b_c)$ is 1.52 m. Figure 3.8 illustrates the transformed 3D layout into the 2D equivalent walls aiming to perform the stability analysis through 2D plane strain condition.

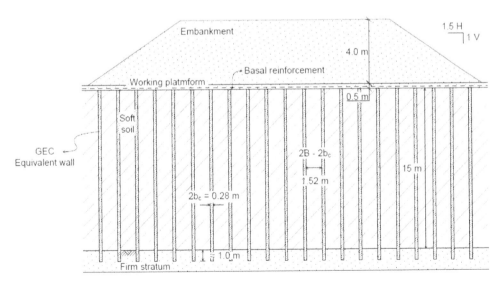

Figure 3.8 Hypothetical model used for the stability analysis with GECs transformed from cylinders to walls

The equivalent column friction angle (ϕ'_{sub}), considering the contribution of encasement, is obtained by means of Equation (2.36) with parameters determined from Equations (2.6) and (2.3).

$$\Delta\sigma_{h,geo} = \frac{F}{r_{geo}} \quad \text{(see Equation (2.6))}$$

$$\Delta\sigma_{h,c} = \Delta\sigma_{v,c}.K_{a,c} + \Delta\sigma_{v,0,c}.K_{a,c} \quad \text{(see Equation (2.3))}$$

For the given encasement tensile force equal to $F = 50$ kN m^{-1} and the column radius $r_c = r_{geo} = 0.40$ m, the encasement confining stress $\Delta\sigma_{h,geo}$ is 125 kPa.

Prior to calculating the column inner horizontal stress $\sigma_{h,c}$, it is necessary to determine the following variables, $\Delta\sigma_{v,c}$, $K_{a,c}$, and $\sigma_{v,0,c}$. The value of $\Delta\sigma_{v,c}$ was already found to be 400 kPa (see chart 12C). The value of $K_{a,c}$ (column active thrust coefficient) is computed using the friction angle of column fill (sand) equal to $\phi'_c = 30°$, thus resulting in $K_{a,c}$ equal to 0.33. The value of $\sigma_{v,0,c}$ (initial vertical stress in the middle of column, depth $z = 7.5$ m), determined by the unit weight of column fill material in a submerged condition, is as follows:

$$\sigma_{v,0,c} = 7.5\times(19-10) = 67.5 \text{ kPa}$$

Using values calculated above and replacing into Equation (2.3) results to a column horizontal stress $\sigma_{h,c}$ of:

$$\sigma_{h,c} = \Delta\sigma_{v,c}.K_{a,c} + \sigma_{v,0,c}.K_{a,c} = (400\times0.33)+(67.5\times0.33) = 155.7 \text{ kPa}$$

Using Equation (2.35) and considering the values of $\Delta\sigma_{h,geo}$ and $\sigma_{h,c}$ and the friction angle of the column fill material $\phi'_c = 30°$, the modified angle of friction ϕ'_{sub} is calculated as follows:

$$\sin\phi'_{sub} = \frac{\left(\dfrac{1+\sin 30°}{1-\sin 30°}\right)+\left(\dfrac{125}{155.7}\right)-1}{\left(\dfrac{1+\sin 30°}{1-\sin 30°}\right)-\left(\dfrac{125}{155.7}\right)+1}$$

$$\sin\phi'_{sub} \cong 0.88 \Rightarrow \phi'_{sub} = 61°$$

The stability analysis was performed based on the 2D transformed geometry (see Fig. 3.9) and increased angle of friction ϕ'_{sub}, considering the contribution of the encasement on the shear strength of column material.

A basal geogrid with maximum mobilized tensile force $T = 60$ kN m^{-1} was placed on top of the encased columns. The encased columns were driven 1.0 m inside the firm stratum (resistant soil). As a practical tip, a 0.5 m thick sand working platform was also considered crossing the encased columns.

Two types of slope stability analyses should be performed. One should consider the short-term ($\phi = 0$) condition using the undrained shear strength profile for the soft soil ($S_u = 14 + 1.0z$). The other should consider the long-term condition, thus taking into account the soft soil drained strength parameters ($c' = 2$ kPa and $\phi' = 25°$). Table 3.3 summarizes the

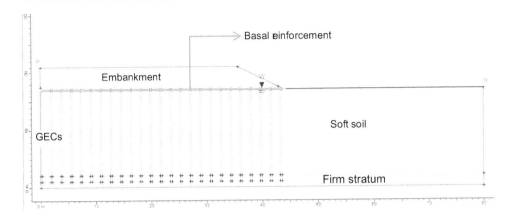

Figure 3.9 Model utilized for slope stability analysis

Table 3.3 Parameters utilized in stability analysis

Material	γ_n (kN m⁻³)	S_u (kPa)	c' (kPa)	ϕ' (°)
Embankment	19	–	2	35
Sand (working platform)	20	–	0	35
Soft soil	14	14 + 1.0z (*)	2 (**)	25 (**)
Firm soil stratum	19	–	10	35
Fill material of GEC	19	–	0	61

(*) short-term undrained conditions
(**) long-term drained conditions

Table 3.4 Results for the slope stability analysis of the embankment

Method of analysis	FS Short-term condition Soft clay: S_u = 14 + 1.0 z		FS Long-term condition Soft clay: c' = 2 kPa and ϕ' = 25°	
	Global failure		Global failure	Localized failure
Janbu Simplified	1.40		2.45	1.80
Bishop Simplified, Spencer, Morgenstern-Price	1.50		2.68–2.69	1.89

FS = Factor of safety.

material parameters used in the stability analysis of the embankment. The friction angle of the GEC ("equivalent wall") was replaced by ϕ'_{sub} determined earlier. The results are presented in Table 3.4 and Figures 3.10 and 3.11.

It can be observed that the column parameter ϕ'_{sub} was found, considering that the system has already reached the equilibrium. Actually, the parameter changes with time until end of consolidation.

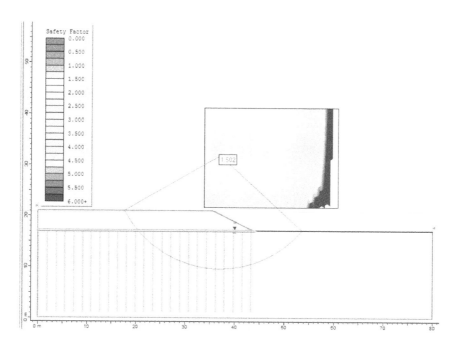

Figure 3.10 Short-term slope stability analysis with basal reinforcement – Method: Morgenstern-Price, FS = 1.50

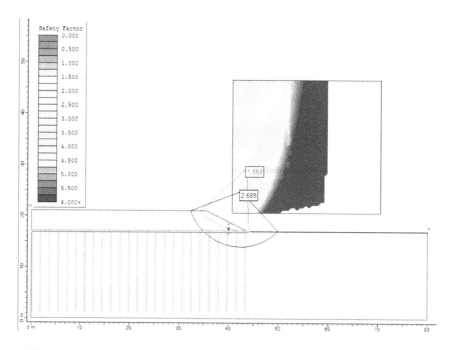

Figure 3.11 Long-term slope stability analysis with basal reinforcement – Method: Morgenstern-Price, FS = 2.69

Some comments on these results are presented below:

a. Factors of safety are greater for long-term conditions, which is typically observed.
b. Factors of safety obtained by circular failure methods, Bishop Simplified, Spencer, and Morgenstern-Price are quite close, also typically observed.
c. The lower factor of safety for long-term conditions may be related to a localized failure circle going through the embankment (see Fig. 3.11 and Table 3.4).
d. The same stiffness and tensile reinforcement values were also used for short- and long-term conditions. However, considering creep effects, this consideration is a simplification of real behavior, in theory these values changes, but it has not affected the present case analysis as the short-term condition prevailed.

3.5.3 Example 3: time-dependent settlement

This example is given in order to estimate the settlement development (S) and the percentage of consolidation (U) *versus* time for the same problem presented in example 2. Following complementary data is assumed to facilitate the calculation process:

- $k_c = 1.0$ m d^{-1} (permeability coefficient of column filling material);
- $k_r = 2.15 \times 10^{-5}$ m d^{-1} (horizontal or radial permeability coefficient of soft soil);
- $k_r/k_s = 2$ (K_s is horizontal permeability coefficient of soft soil in smear zone);
- $d_s/d_c = 2$ (d_s is the diameter of smeared zone and d_c is the column diameter);
- $c_h = c_r = 3.0 \times 10^{-8}$ m^2 s^{-1} (soft soil horizontal or radial coefficient of consolidation).

Solution

The analytical method proposed by Han and Ye (2002), described in Chapter 2, is used to solve the present problem. Equation (2.24) is used to calculate the dimensionless value of F'_m. For this purpose, the following variables should be initially introduced: N, S_{mr}, k_r, k_s, k_c, H_d, and d_c.

Where:
$N = d_e/d_c = 2.03/0.80 = 2.54$;
$S_{mr} = d_s/d_c = 2$;
$k_r = 2.15 \times 10^{-5}$ m d^{-1} ;
$k_s = 1.75 \times 10^{-5}$ m d^{-1} ;
$k_c = 1$ m d^{-1} ;
$H_d = 7.5$ m (= 15/2, half of the thickness of soft soil);
$d_c = 0.80$ m.

Substituting the above parameters into Equation (2.24), the value of F'_m is equal to 0.769. The next step is to calculate the value of T_{rm} by Equation (2.22) as a function of the time t. In order to compute the value of T_{rm}, the modified radial consolidation coefficient (c_{rm}) should be initially computed using Equation (2.23) considering the following values:

$k_r = 2.15 \times 10^{-5}$ m d^{-1};
$c_h = c_r = 3 \times 10^{-8}$ m^2 s^{-1};

N = 2.54;

$n_s = \Delta\sigma_{v,c}/\Delta\sigma_{v,s} = 400/20 = 20$

Replacing the values above into Equation (2.23) results in a c_{rm} value equal to 1.4×10^{-7} m² s⁻¹. Therefore, the value of T_{rm} can be determined as a function of time t. Finally, the percentage of consolidation (U%) at a specific time t is determined by Equation (2.21). Based on the calculated U, the settlement-time curve is plotted using Equation (2.20) as follows:

$$S(t) = S_c.U(t)$$

Where S_c is the settlement value computed according to Raithel and Kempfert's (2000) method. Since the settlement computed in example 2 is $S_c = S_s = 0.39$ m, this value will be used to find the settlement S(t) at any desired time during consolidation. For the case studied, the settlement-time curve is illustrated in Figure 3.12. It is observed that approximately 3.3 months (100 days) are required to reach 95% of consolidation with the maximum embankment settlement of about 0.4 m.

3.5.4 Example 4: lateral thrust on piles

Background

The increase in soil vertical stress ($\Delta\sigma_{v,s}$) due to embankment loading generates an increase in horizontal stresses ($\Delta\sigma_{h,s}$), which are subsequently transferred into the adjacent structures (Tschebotarioff, 1962, 1970, 1973). These horizontal stresses may impose excessive lateral

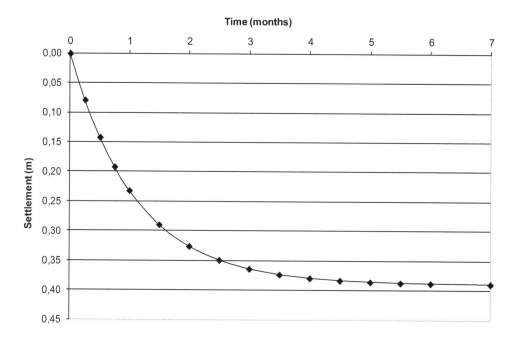

Figure 3.12 Example 3: Settlement curve *versus* time

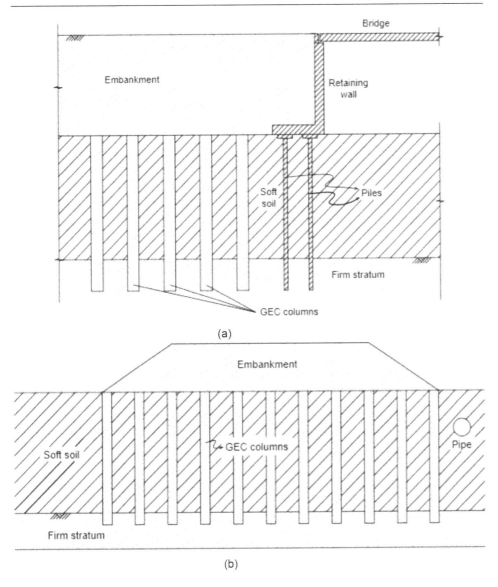

Figure 3.13 Examples where horizontal thrust should be considered (Riccio et al., 2012, 2016)

deformation on these structures, consequently causing structural damages or local failure. GECs installed in the soft soil below, e.g., an embankment reduce these horizontal stresses, as the columns absorb most of embankment stresses. As shown in Figure 3.13, an additional lateral thrust arises when the load on the soft soil is asymmetric in relation to a buried element. Since a GEC foundation system generally reduces the vertical stress on soil $\Delta\sigma_{v,s}$, as shown above, it consequently reduces the lateral thrust mentioned.

In these cases, it is necessary to calculate that portion of the embankment applied load supported by the encased column ($\Delta\sigma_{v,c}$) and the part acting on the soil ($\Delta\sigma_{v,s}$). As discussed in section 2.3, the value of $\Delta s_{v,s}$ can be calculated based on Raithel and Kempfert's (2000) analytical method.

Almeida *et al.* (2014) measured the stress concentration factor ($n_s = \Delta\sigma_{v,c}/\Delta\sigma_{v,s}$) for a test embankment over group of GECs. It was observed that the values of n_s measured below the embankment centerline varied between 2.2 and 2.4, but values of n_s equal to 5 or higher have been reported (e.g., Raithel *et al.*, 2012).

Several studies (e.g., Murugesan and Rajagopal, 2006; Riccio *et al.*, 2012; Hosseinpour, 2015) have shown that the use of GECs reduces significantly the vertical and horizontal stresses in the soft soil layer, since they support most of the applied vertical stresses. The reduction of $\Delta\sigma_{h,s}$ (horizontal stress in the soil) is proportionally related to the reduction in the value of $\Delta\sigma_{v,s}$ (soil vertical stress).

Tschebotarioff (1962) empirically quantifies the increase in horizontal stress $\Delta\sigma_{h,s}$ by Equation (3.1):

$$\Delta\sigma_{h,s} = K.\Delta\sigma_{v,s} \qquad\qquad (3.1)$$

where:

$\Delta\sigma_{v,s}$: increase in vertical stress on soft soil;
K: coefficient of thrust, usually assumed as $K = 0.40$, according to Tschebotarioff (1970, 1973).

Tschebotarioff (1970, 1973) used field measurements and defined the maximum value of the lateral force acting per meter along the pile, which is equal to:

$$P_h = B_L \Delta\sigma_{h,s} \qquad\qquad (3.2)$$

Where B_L is the width of the adjacent pile.

The distribution of P_h along the pile is assumed to be triangular as shown in Figure 3.14. The top rigid soil in Figure 3.14 represents a sand layer or a working platform layer placed above the soft soil.

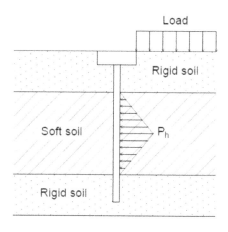

Figure 3.14 Distribution of the lateral force along the pile in soft soil (Tschebotarioff, 1970, 1973)

Another formulation to evaluate the horizontal force acting on the pile, P_h, is provided by De Beer and Wallays (1972). This approach is recommended only when the global factor of safety, neglecting the piles in the stability analysis, is higher than 1.6. According to this approach, the value of $\sigma_{h,p}$ (horizontal stress) acting on the pile is given by Equation (3.3), considering a constant load distribution along depth, as follows:

$$\sigma_{h,p} = q \qquad (3.3)$$

Where q is the applied vertical load. The horizontal stress on the pile, considering the use of GECs columns, is given by Equation (3.4).

$$\sigma_{h,p} = \Delta\sigma_{v,s} \qquad (3.4)$$

Where $\Delta\sigma_{v,s}$ is the increase in the vertical stress on soil considering the presence of GECs columns, i.e., $\Delta\sigma_{v,s}$ is lower than the applied load q.

The horizontal force per meter on the pile can be then estimated considering the use of GEC, by means of Equation (3.5).

$$P_h = B_L \Delta\sigma_{h,p} \qquad (3.5)$$

Where $\Delta\sigma_{h,p}$ is the horizontal stress on soil due to the load.

Recently, GECs have been successfully used in some practical applications with the purpose of reduction of the lateral thrust on piles (Alexiew et al., 2016; Schnaid et al., 2017). Following, an example is given aiming to determine the lateral thrust on piles supporting bridge abutment when GECs is utilized.

Practical example

A 7 m high bridge abutment is considered over a 10-m thick soft clayey foundation with the average constrained modulus ($E_{oed,ref}$) equal to 500 kPa and shear strength parameters $c' = 2$ kPa and $\phi' = 25°$. The water table is coincident with the top of the soft soil layer. The bridge abutment consists of a reinforced concrete wall over piles. The unit weight of the compacted wall fill is 19 kN m^{-3}. The bridge foundation piles near the abutment base are under impact of the horizontal forces induced by the horizontal thrust in the soft soil. This horizontal thrust is caused by the increase of the vertical stress on soil ($\Delta\sigma_{v,s}$) due to the load applied by the embankment ($\Delta\sigma_0$). The circular bridge piles are 0.60 m in diameter. The abutment is founded on 0.80 m diameter GECs, as schematically shown in Figure 3.15. The columns are installed in a square arrangement with an average center-to-center spacing of s = 2 m while encased by the geosynthetic with tensile stiffness of J = 3500 kN m^{-1}.

Solution

Initially, the total stress imposed by the bridge abutment is computed as:

$$\Delta\sigma_0 = 7 \, m \times 19 \, kN / m^3 = 133 \, kPa$$

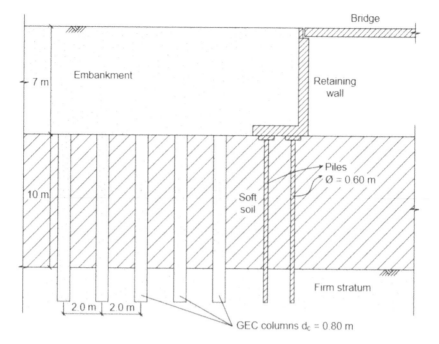

Figure 3.15 Hypothetical example of bridge abutment over GECs improved foundation

The area replacement ratio a_E (A_c/A_E), based on the GECs pattern, is also calculated as follows:

$d_e = 1.13s = 2.26$ m
$A_E = 4.01$ m²
$d_c = 0.80$ m
$A_c = 0.50$ m²
$a_E = A_c/A_E = 0.125$ ($a_E = 12.5\%$).

Since the pre-design charts are not calibrated for intermediate a_E values (e.g., $a_E = 12.5\%$), the curve corresponding to $a_E = 10\%$ is conservatively used to obtain the vertical stress on soil. The horizontal force on piles can be then calculated using Equations (3.1) and (3.2), according to Tschebotarioff's (1962) method, as mentioned earlier.

According to this approach, the empirical coefficient K is assumed to be 0.40. The chart 7A (Annex I) is then utilized to compute the value of $\Delta\sigma_{v,s}$ considering the soil parameters given. Based on Figure 3.16, the value of the increase in vertical stress on soft soil due to the load imposed by the bridge abutment is about 15 kPa.

Following Tschebotarioff's (1962) method, the horizontal stress in soil ($\Delta\sigma_{h,s}$) is given by:

$$\Delta\sigma_{h,s} = K.\Delta\sigma_{v,s} = 0.4 \times 15 = 6 \text{ kPa}$$

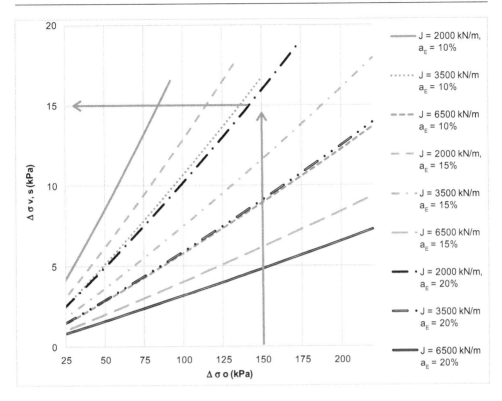

Figure 3.16 Chart 7A (Annex I) – determination of $\Delta\sigma_{v,s}$

Thus, the force per meter acting on the pile is calculated considering B equal to the pile diameter:

$$P_h = B_L \cdot \Delta\sigma_{h,s} = 0.6 \times 6 = 3.6 \text{ kN m}^{-1}$$

It is important to mention that, as expected, the horizontal force on piles P_h is much lower than the force that would act on the piles would be if there were no GECs. This is because the GECs reduce the increase of vertical stress on soil ($\Delta\sigma_{v,s}$) caused by the embankment loading ($\Delta\sigma_0$). In the absence of GECs, the force acting on piles would be:

$$P_h = B_L \cdot K \cdot \Delta\sigma_0 = 0.6 \times 0.4 \times 133 = 31.9 \text{ kN m}^{-1}$$

If the force per meter is supposed to be calculated by De Beer and Wallays's (1972) solution, the value of P_h is given by $P_h = B_L \cdot \Delta\sigma_{v,s} (= B_L \cdot \Delta\sigma_{h,p})$. It is noted that, $q = \Delta\sigma_0 (= 133 \text{ kPa})$ in the absence of GEC columns and $q = \Delta\sigma_{v,s} = \Delta\sigma_{h,p} (= 15 \text{ kPa})$ with the presence of GEC columns in the mentioned arrangement. So, the force per meter on the pile using De Beer and Wallays's (1972) solution is $P_h = B_L \cdot \Delta\sigma_{v,s} = 0,6 \times 9 = 9 \text{ kN m}^{-1}$. Therefore, the value of P_h considering the presence of GEC columns and using De Beer and Wallays's (1972) method is 9 kN m^{-1}, therefore about 150% higher than Tschebotarioff's (1962) method.

Table 3.5 Impact on the piles considering the absence of GEC columns and presence in GEC columns

Horizontal stress/ horizontal force	Without GEC columns (using K = 0.4)	With GEC columns (Tschebotarioff)	With GEC columns (De Beer and Wallays)
$\Delta\sigma_{h,s}$ (kPa)	53.2	6	15
P_h (kN m^{-1})	31.9	3.6	9

Table 3.5 summarizes the values of $\Delta s_{h,s}$ (increase in horizontal stress on soil) and P_h (force per meter in the piles) considering the following conditions: (a) without GEC columns; (b) with GEC columns and considering Tschebotarioff's (1962) method, and (c) with GEC columns and considering De Beer and Wallays's (1972) method.

Chapter 4

Instrumented embankments on GEC

4.1 Introduction

The geotextile encased column (GEC) foundation system for embankments and dikes on soft soil deposits were introduced more than 25 years ago in Germany and since then several projects have been successfully executed worldwide. A number of calculation methods are available (see Chapter 2) and a design procedure is recommended by EBGEO (2011), and two methods of GEC installation (see Chapter 1) have been extensively tested in practice.

However, for better understanding of the behavior of the GEC system, the field response of a trial embankment constructed on GECs-stabilized soft foundation is described with details in this chapter. Afterwards, some other important case studies and practical projects on which GECs were utilized are briefly mentioned.

4.2 Test embankment at TKCSA, Itaguaí, Brazil, 2012

In 2012, a 5.3 m high test embankment was built at a test area in the stockyard of ThyssenKrupp Company, located in Itaguaí, state of Rio de Janeiro, Brazil. The stockyard was an operating area (covering almost 0.5 km²) for the storage of the raw coal, ore, and additives used in the production of steel. The soil profile at the stockyard consisted of a very soft clay layer that extended from near the ground surface to a depth ranging from 9 to 10 m (see Fig. 4.1).

Thirty-six geotextile encased granular columns, installed in a 6 × 6 grid, strengthened the test area. The columns were 80 cm in diameter and 11 m in length, encased by seamless woven geotextile (named Ringtrac 100/250), and implemented on 2 m center-to-center spacing on a square pattern. According to the site investigation performed, the soft clay was classified as CH with an average undrained strength (S_u) of about 15 kPa. The soft clay properties obtained from site investigation are presented in Table 4.1. The column filling material was poorly graded crushed stone with minimum and maximum particle sizes of 10 mm and 35 mm, respectively. It is noted that the encased columns were installed in 2008, but the field test was performed in 2012. This four-year period was more than enough to dissipate the effect (pore pressures) of column installation on the behavior of the test embankment.

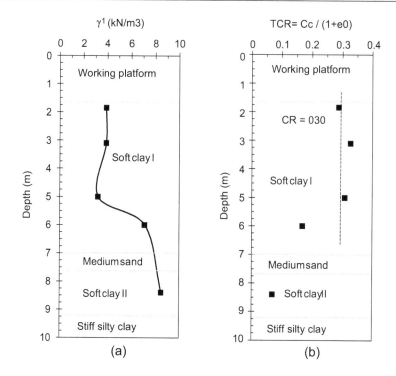

Figure 4.1 Soft clay properties: (a) submerged unit weight; (b) compressibility ratio by oedometer tests

Table 4.1 Geotechnical properties of TKCSA soft clay

Depth (m)	γ'	C_c	C_s	e_0	$k_v \times 10^{-10}$	$k_h \times 10^{-10}$	$c_v \times 10^{-8}$	$c_h \times 10^{-8}$	ϕ'	c'
	$(kN\ m^{-3})$	$(-)$	$(-)$	$(-)$	$(m\ s^{-1})$	$(m\ s^{-1})$	$(m^2\ s^{-1})$	$(m^2\ s^{-1})$	$(^\circ)$	(kPa)
2.85–3.35	4.2	1.25	0.92	2.85	1.03	1.72	3.09	5.14	27.6	5.0
5.75–6.25	4.3	0.38	0.05	1.23	0.567	0.785	2.05	2.84	25.4	0.0
8.15–8.65	7.2	0.13	0.04	0.87	0.565	0.878	3.17	0.437	28.6	3.0
1.60–2.10	5.0	1.03	0.07	2.60	1.91	3.93	1.85	3.80	27.3	0.0
4.75–5.25	4.2	1.27	0.13	3.16	1.44	4.86	4.36	0.147	22.8	1.0

4.2.1 Instrumentation and embankment construction

Figure 4.2 shows the sectional view of the embankment and the position of the instrumenta-tion. A biaxial geogrid was placed prior to the embankment construction to follow EBGEO's (2011) overall recommendations to use basal reinforcement under circumstances of high embankment loads. As shown in Figure 4.2, the soft soil and encased columns were instru-mented to register vertical displacement (surface settlement plate, S), horizontal deformation (inclinometer, IN), pore water pressure (vibrating wire piezometer, PZ), vertical stresses (total stress cell, CP), and geotextile hoop strain (radial extensometer, EX). Details of the instrumentation are presented in Table 4.2.

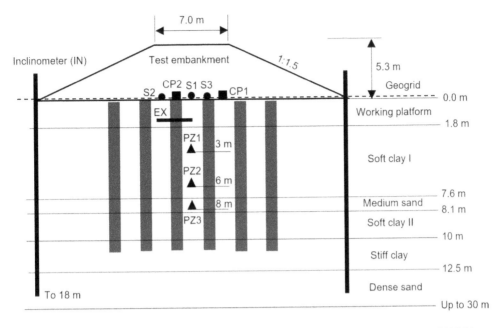

Figure 4.2 TKCSA embankment sectional view and instrumentation position (Hosseinpour, 2015; Hosseinpour *et al.*, 2017)

Table 4.2 Summary of the instrumentation used in TKCSA test embankment (Hosseinpour, 2015)

Type of instruments	Quantity	Location of instruments	Purpose
Total stress cell (CP)	4	On the top of the encased columns and between them	Measure the vertical stresses on top of the encased column and surrounding soft soil.
Piezometer (PZ)	3	Installed 3 m, 6 m, and 8 m below ground surface in soft clay layer	Measure pore pressure dissipation and consolidation process.
Settlement sensor (S)	3	On the top of the encased columns and between them	Measure the settlements below the embankment. Allows for evaluation of differential settlements.
Diameter extensometer (EX)	3	Attached to the geotextile encasement at 1 m below column top	Measure column bulging. Allows evaluation of average ring strains and tensile force.
Inclinometer (IN)	2	Installed at embankment toes	Measure the horizontal deformation beneath the embankment toes.
Data logger	1	16 channels	Data storage and collection.
LAB-III multi-stage surge protection	13	–	Protection against electrical discharge for each instrument.

Figure 4.3 Typical moments of TKCSA test embankment construction (Hosseinpour, 2015)

The material used for the test embankment was sinter feed obtained during the ore enrichment process. The sinter feed, classified as SW, had a friction angle of 35° measured by direct shear test. The embankment construction was performed in four layers including the consolidation interval between them. The average unit weight of the fill material and the natural water content were 27.8 kN m⁻³ and 6.6%, respectively. The final height of the embankment was 5.35 m, producing almost150 kPa of total vertical stress. The embankment was left in place for 180 days after the final layer was placed, at which time most of the excess pore pressure had already dissipated. In Figure 4.3, some typical moments of test embankment construction are illustrated.

4.2.2 Vertical and horizontal deformations

The settlement measured on the top of the encased columns and between them are plotted in Figure 4.4. The settlements increased in the construction stages when embankment height increased and also in the post-construction period, when excess pore pressure dissipated. The maximum settlement occurred at the midpoint between the encased columns.

The results of TKCSA test embankment were also compared (Hosseinpour *et al.*, 2016) with an embankment (Magnani, 2006; Magnani *et al.*, 2009) built over unstrengthened soft soil. Both the test embankment and the soft soil below were quite similar in terms of geo-metrical and geotechnical parameters. It is seen that the GEC system reduces the settlements significantly. In addition, the GECs increased the load-carrying capacity of the soft founda-tion. This fact provides an example of ground improvement benefits regarding the bearing performance of the soft foundation.

Figure 4.5 shows the profile of the soil horizontal deformations measured underneath the embankment toes during construction. The maximum horizontal displacement increased when embankment height increased, as expected. It is also seen that the maximum soil hori-zontal displacement occurred at the middle of soft clay, confirmed by both inclinometers on which depth where the collected samples had shown the lowest constrained stiffness modulus of soft soil.

The horizontal deformation of TKSCA test embankment is compared with conven-tional embankment (Magnani, 2006) in Figure 4.6, and it is seen that the GECs reduced

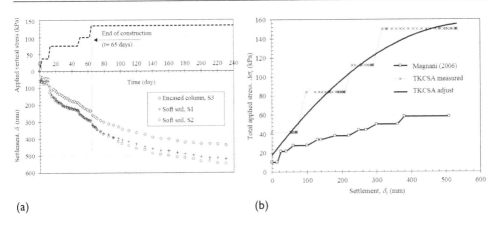

(a)   (b)

Figure 4.4 (a) Settlement development *vs.* time in TKCSA project (Almeida *et al.*, 2014; Hosseinpour, 2015); (b) effect of GECs on embankment settlement (Hosseinpour *et al.*, 2016)

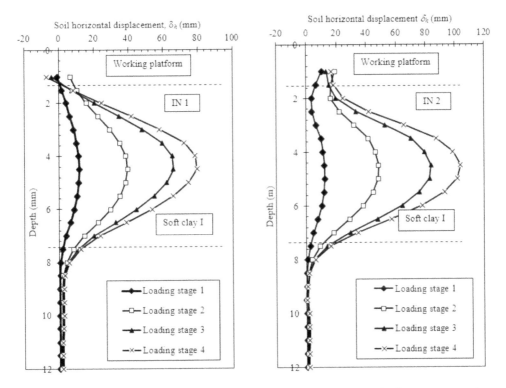

Figure 4.5 Profile of the soil horizontal deformation measured during embankment construction (Almeida *et al.*, 2014; Hosseinpour, 2015)

significantly the maximum horizontal displacement of the soft foundation. In other words, the use of GECs notably enhanced the embankment stability against the failure occurring due to large horizontal displacement of the soft foundation. This confirms the similar finding regarding a significant increase of embankment global stability in Raithel *et al.* (2012).

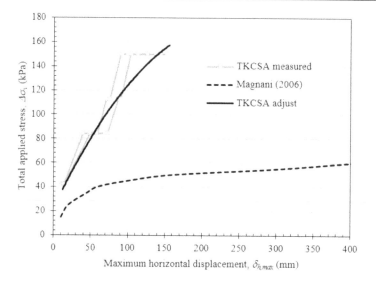

Figure 4.6 Effect of GECs on soil horizontal deformation (Hosseinpour et al., 2016)

Maximum horizontal displacement of the foundation soil is correlated with the maximum embankment settlement, measured by sensor S1, as shown in Figure 4.7. It is common (e.g., Tavenas et al., 1979) to analyze the ratio between these two measurements through:

$$DR = \frac{\delta_{h,max}}{\delta_{v,max}}$$
(4.1)

From Figure 4.7, it can be seen that the horizontal displacement increases about linearly with settlement resulting in a slope (DR) varying from 0.16 to 0.20. From the analyses of fifteen embankments built on soft deposits without ground improvement, Tavenas et al. (1979) reported average DR ratios equal to 0.91 at yield conditions during construction and equal to 0.16 during consolidation. It can be concluded that GECs resulted in roughly uniform DR values during the construction and consolidation stages, unlike conventional embankments on nonimproved foundations. Comparison with similar conventional embankment (Magnani, 2006) indicates that the GECs caused the maximum horizontal deformation increased with quite slower DR of about 4 times less than the ordinary embankment.

4.2.3 Vertical stresses below the embankment

The development of the total vertical stresses measured on the top of the encased columns and between them are plotted in Figure 4.8. The plausibility of the measured vertical stress can be assessed by checking the vertical load equilibrium between embankment applied stress ($\Delta\sigma_v$) versus total vertical stresses carried by the encased column ($\Delta\sigma_{v,c}$) and the surrounding soft soil ($\Delta\sigma_{v,s}$) (Aboshi et al., 1979), as follows:

$$\Delta\sigma_v \times A_E = \Delta\sigma_{v,c} \times A_c + \Delta\sigma_{v,s} \times (A_E - A_c)$$
(4.2)

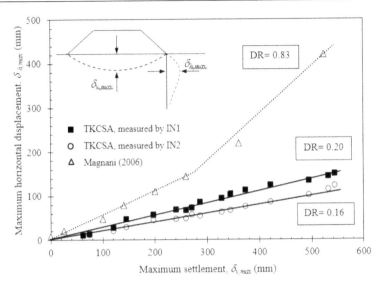

Figure 4.7 Relation between maximum embankment settlement and maximum horizontal deformation (Hosseinpour *et al.*, 2016)

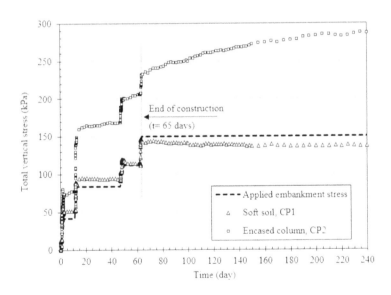

Figure 4.8 Development of total vertical stress below the TKCSA test embankment (Almeida *et al.*, 2014; Hosseinpour, 2015)

Where A_c and A_E represent respectively the cross-section area of the column and the unit cell determined by the column pattern. Based on Figure 4.8, the final total vertical stresses supported by column and surrounding soft soil are 280 and 135 kPa, respectively. Using Equation (4.2) results in a total vertical load equal to 612.5 kN transmitted to the top of the unit cell, which is sufficiently close (error 2%) to the total vertical load applied by the embankment ($\approx$ 150 kPa). The greater stiffness of the encased column, the higher vertical stress is transferred to the GEC resulting in a stress concentration factor of about 2.4 at the end of monitoring time. This value, however, is lower than usually reported in other similar cases (e.g., Raithel *et al.*, 2012). A possible explanation could be due to additional stress redistribution induced by the relatively thick working platform up to the level of the pressure cells.

4.2.4 Excess pore water pressure

Excess pore water pressures measured by piezometers at depths of $z = 3$ m, 6 m, and 8 m are shown in Figure 4.9a. The excess pore pressures tend to increase sharply just after each layer of the embankment was placed because of quick loading conditions induced to the soft clay, followed by a gradual dissipation during post-construction.

In Figure 4.9b, the excess pore pressure measured in middle of soft clay layer is compared with similar conventional test embankment (Magnani, 2006). It is seen that the GEC system caused the maximum excess pore pressure to reduce significantly, while its total vertical load was 2.5 times greater. The reason is found to be due to stress concentration on the top of the GEC resulting in less total vertical load transferred to the soft clay foundation. Considering the dissipation time, the radial drainage offered by the granular columns working also as drains with very high drainage capacity led the consolidation time to speed up significantly.

4.2.5 Geotextile expansion (column bulging)

Development of the geotextile hoop strain measured by different extensometers attached to the geotextile at the depth 1.0 m below the column top is shown in Figure 4.10a. It can be seen that the hoop (ring) strains increased during loading stages as embankment height increased as well as during consolidation when the excess pore pressure is dissipated. The delayed expansions of the geotextile encasements following loading stage 4 is in accordance with the overall stress-strain behavior of the columns observed by the continuous column settlement and vertical stresses. The further geotextile expansion during consolidation caused the column to sustain greater total vertical stress and thus stress concentration

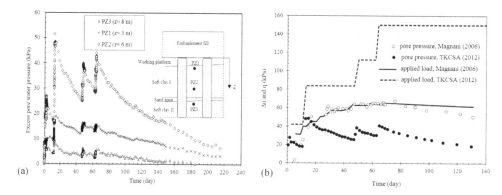

Figure 4.9 (a) Excess pore pressure development *vs.* time for TKCSA embankment; (b) effect of GECs on variations of excess pore pressure (Hosseinpour *et al.*, 2016)

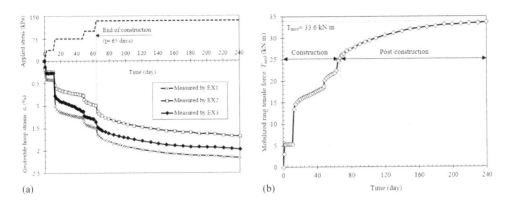

Figure 4.10 Geotextile radial deformation: (a) development of hoop strain; (b) average mobilized ring tensile force in geotextile encasement (Almeida et al., 2014; Hosseinpour, 2015)

developed at a slow rate. The mobilized ring force of the geotextile T_{mob} can then be calculated by hoop strain measurements as follows:

$$T_{mob} = \frac{\Delta d_c}{d_c}.J \qquad (4.3)$$

Where d_c and J are the column original diameter and geotextile stiffness modulus respectively equal to $d_c = 80$ cm and J = 1750 kN m^{-1}.

According to Figure 4.10b, the mobilized ring tensile force increased by load applications followed by a continuous increase during post-construction. It is also seen that the mobilized ring tensile force at the end of monitoring time is equal to 33.6 kN m^{-1}, which is around 35% of the maximum ring tensile force (available after application of reduction factors) of the geotextile encasement (i.e., 95 kN m^{-1}). It should be noted, however, that the lateral constraint of the GECs in the sand working platform, where the bulging was measured (Fig. 4.2), is higher than in the soft soil below, thus strains and tensile force in the zone below the platform can be higher.

4.3 Other case studies

The following items present briefly the results of applying GECs in practice. Many of the cited projects were performed with Dr. M. Raithel from Kempfert + Raithel Geotechnik, Germany, as design engineer. More detailed information can be found in Alexiew and Raithel (2015).

4.3.1 Railroad embankment at Waltershof, 1995

Near Waltershof, a heavy loaded railroad to the harbor of Hamburg – positioned on a 5 m high embankment – had to be widened due to increasing traffic. The subsoil consists of about 5 to 6 m of soft saturated clays and peat. The existing 5 m embankment had settled over the years by 1.2 to 1.5 m. There were two lots in the project. Due to logistic reasons, Lot 1 had to be executed in a month and had a maximum time allowed for consolidation of four months. So, it was decided to build the new embankment in Lot 1 on GECs.

As shown in Figure 4.11, the GECs had a diameter of 1.54 m; the area ratio was in the range of 20–30%. The encasement was produced from a geocomposite Comtrac 200/50 B 30 with an ultimate tensile strength (UTS) in the ring direction of 200 kN m^{-1} and an average ring tensile modulus of J = 1800 kN m^{-1}. The typical average soil parameters are given in Table 4.3.

A measurement program was implemented comprising horizontal and vertical incli-nometers and earth pressure cells (both over and between the GECs) and also piezometers

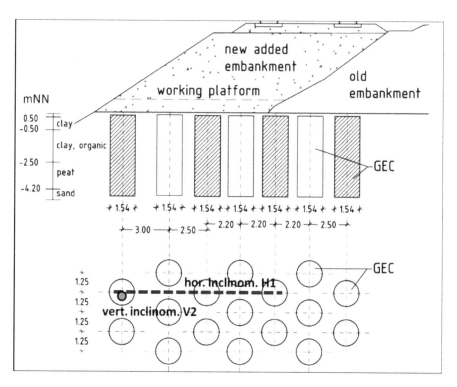

Figure 4.11 Typical scheme of the railroad embankment on GEC at Waltershof (Alexiew and Raithel, 2015)

Table 4.3 Average soil parameters of the subsoil layers at Waltershof (Alexiew and Raithel, 2015)

Soil layer	Position	γ/γ_{sat}	k	E_{oed}	ϕ'	c'
	(m)	(kN m^{-3})	(m s^{-1})	(MN m^{-2})	(°)	(kPa)
Upper clay	+0.5 to 0.5	19/19	1×10^{-9}	2.6	29.0	8.0
Clay, organic	−0.5 to 2.5	13/13	1.5×10^{-8}	0.8	25.5	16.0
Peat	−2.5 to 4.2	11/11	1.4×10^{-7}	0.6	20.5	8.5
Sand	−4.2	19/20	3×10^{-5}	27.0	35.0	1.0

γ/γ_{sat}: unit weight; natural/saturated; k: coefficient of permeability; E$_{oed}$: constrained compression modulus; ϕ': drained angle of internal friction; c': drained cohesion.

in the soft soils. In Figure 4.12a, typical horizontal displacements of a first GEC at the toe of the new embankment are depicted, for example, in a zone of generally intensive "spreading" of embankments on soft soils (see Fig. 4.11, V2); the final horizontal displacements of the toe amount to 9 cm, which is practically negligible, under the given conditions. Figure 4.12b shows the settlements over a row of GECs across the embankment (see Fig. 4.11, H1); note the missing support by GEC at 8 m from the toe (left edge) under full height embankment resulting in significantly larger local settlement.

Figure 4.13a provides time-dependent data on load development and corresponding settlements on top of the GECs and on top of the soft soil. The settlements between the

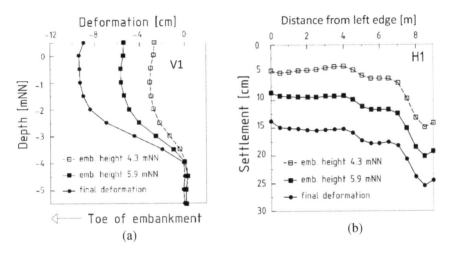

Figure 4.12 (a) Horizontal displacements of vertical inclinometer V1; (b) settlements of horizontal inclinometer H1 (Alexiew and Raithel, 2015)

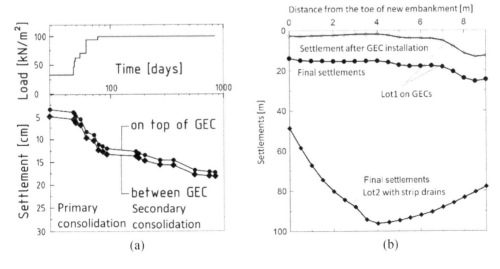

Figure 4.13 (a) Typical settlements in Lot 1 (on GECs); (b) comparison of settlements between Lot 1 and Lot 2 (Alexiew and Raithel, 2015)

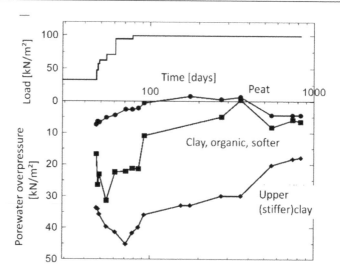

Figure 4.14 Typical development of pore water overpressure over time (Alexiew and Raithel, 2015)

GECs are a bit larger. This is controversial, however, due to the formal assumption of equal settlements in Raithel (1999); Raithel and Kempfert (2000); and EBGEO (2011). Figure 4.13b shows an interesting comparison of settlements across the new embankment between Lot 1 on GECs and the unsupported Lot 2 with strip drains only. Note that geometry, subsoil, loads, and other variables for both are practically identical; the only difference is that the construction period for Lot 2 lasted some months longer due to consolidation intervals. It is clearly seen that the embankment over GECs underwent quite lower settlement compared with the embankment over strip drains only.

The vertical stress on top of GECs and between them provided by the earth pressure cells resulted in a column efficiency E (i.e., load supported by GECs/total load applied), varying from 0.4 to 0.6 over the time of construction.

The development of excess pore pressure is plotted in Figure 4.14. Three facts seem to be of interest: the maximum value in the upper clay is less than half of the applied load; although the organic clay is in a disadvantageous position regarding draining, it consolidates quickly; and the pore pressures in all soil layers start to decrease even under a still increasing load before the end of construction. The most plausible explanation is the huge draining capacity of the encased sand columns.

4.3.2 Extension of Airbus site at "Mühlenberger Loch", Hamburg, 2000–2002

The plant site of the Airbus Company in Hamburg-Finkenwerder at the river Elbe was enlarged by about 140 ha for new branches of production, in particular for the production of the new Airbus A380. The extension was carried out by enclosing an area of extremely (8 to 14 m) thick soft soils with undrained shear strength S_u of only 0.4 to 10.0 kPa. The characteristic drained consolidated soil parameters are summarized in Table 4.4.

The dyke was founded on about 60,000 geotextile encased sand columns of corresponding length with a diameter of 80 cm and a total installed length of about 650 km. This is the biggest single GEC job executed so far. A typical cross-section with the GECs can be seen in Figure 4.15.

Table 4.4 Average soil parameters of the subsoil layers at Airbus site Hamburg "Mühlenberger Loch" (Alexiew and Raithel, 2015)

Soil layer	γ/γ_{sat} (kN m^{-3})	k (m s^{-1})	E_{eod} (MN m^{-2})	ϕ' (°)	c' (kPa)
Clay	6/16	2×10^{-10}	0.60	20	0
Sludge	4/14	2×10^{-10}	0.45	20	0
Peat	1/11	1×10^{-8}	0.55	20	0

γ/γ_{sat}: unit weight; natural/saturated; k: coefficient of permeability; E_{oed}: constrained compression modulus; ϕ': drained angle of internal friction; c': drained cohesion.

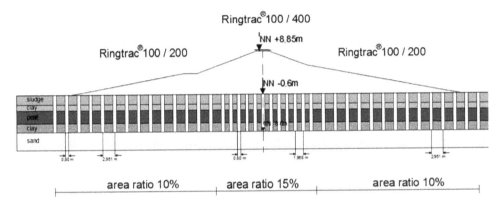

Figure 4.15 Typical dyke cross-section at Airbus site Hamburg "Mühlenberger Loch" (Alexiew and Raithel, 2015)

Figure 4.16 Typical moments of construction at Airbus site Hamburg, overview (Alexiew and Raithel, 2015)

A geocomposite with an ultimate tensile strength of 500–1000 kN m^{-1} was installed on top of GECs combining high-strength and filter stability to accelerate dyke construction and to guarantee global dyke stability, especially in the initial stage of construction. In Figure 4.16, typical moments of construction are depicted.

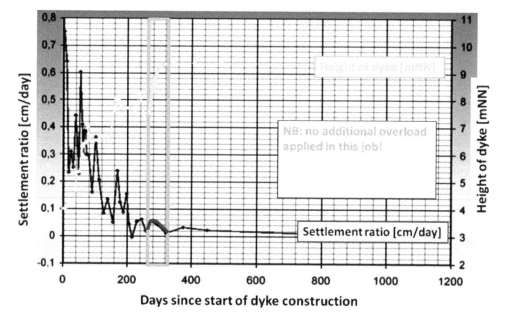

Figure 4.17 Example of measured settlement ratio *vs.* dyke height (Alexiew and Raithel, 2015)

The stability and deformation predictions were verified by on-site measurements during construction. Figure 4.17 shows an example of development of settlements *versus* dyke height and time. It can be seen again that the tendency of the settlement ratio decreases with time despite the increasing height of dyke. But the most interesting point occurs at about day 300: although the dyke height increases after that by two meters, the settlement ratio drops down to almost zero. A possible explanation is the high degree of mobilization of GECs: higher, then predicted by design assumptions.

4.3.3 Railroad embankment Botnia line, Sweden, 2001–2002

The Botnia Line (Botniabanan) was a 190 km long, high-speed railway line in northern Sweden. At Bridge 4, the Botnia line crosses a valley with very soft soils with a depth of up to 7.5 m with an average undrained strength of $S_u = 20$ kPa. The embankments on both sides of the bridge had a height of 9–10 m. Due to the settlement requirements, a GEC foundation with a diameter of 80 cm using the displacement method was carried out in 2001. The GECs were arranged in a triangular pattern with 15% area replacement ratio as shown in Figure 4.18.

The design resulted in the choice of geotextile encasement with an ultimate tensile strength of 400 kN m^{-1} in the ring direction. Crushed fill was used to form the GEC. A significantly higher damage of geotextile encasement during installation and compaction had to be expected in comparison to the experience with sands. For determination of the reduction factor for installation damage, the installation process was simulated and an average reduction factor of 1.36 was evaluated (compared to 1.05 to 1.15 typical for sands) and applied for estimation of the "ring" design strength.

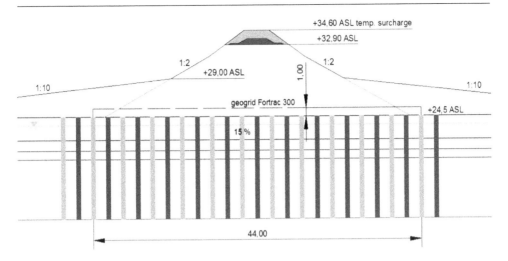

Figure 4.18 Cross-section of the embankment with the GEC-column pattern (Alexiew and Raithel, 2015)

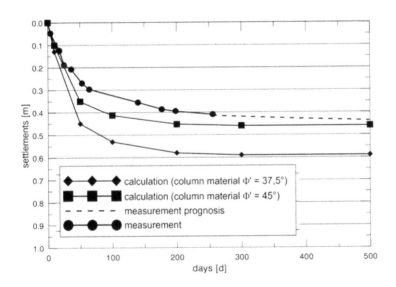

Figure 4.19 Botnia line: calculated and measured settlements (Alexiew and Raithel, 2015)

The design resulted in calculated settlements of about 60 cm. On the safe side, an angle of internal friction for the column fill $\phi' = 37.5°$ was assumed. After construction, the settlements of the embankment were measured for a period of about 250 days (Fig. 4.19), showing significantly smaller settlements. A post-measurement design simulation demonstrates that a $\phi' = 45°$ for the crushed fill is more realistic.

4.3.4 High-speed rail link Paris-Amsterdam, Westrik, Netherlands, 2002

Near Breda, the new high-speed railroad from Paris to Amsterdam had to cross the former waste disposal at Westrik. The waste thickness was in the range of 4–6 m, followed by sands with some thin clay interlayers (Fig. 4.20) which were improved with GECs preferred for ecological, financial, and logistic reasons. The GECs had a diameter of 0.8 m with an average replacement ratio of 15% and a "ring" ultimate tensile strength of 300 and 400 kN m^{-1}.

A total of 2200 GECs were installed in June and July 2002 using sand as fill. The displacement method was applied successfully despite the problematic character of waste. A measurement program was applied using flexible horizontal inclinometers at the top of GECs to record settlement development and to decide when to start installing track on top of the embankment. Results for the left part of a typical cross-section are shown in Figure 4.21. The

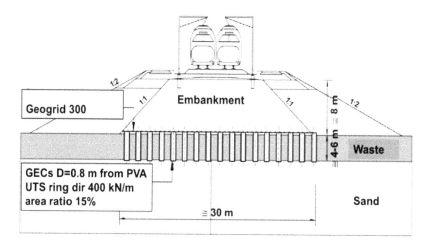

Figure 4.20 HSL Paris-Amsterdam at Westrik: typical cross-section (Alexiew and Raithel, 2015)

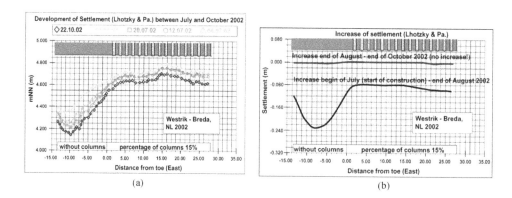

Figure 4.21 (a) Development of settlements; (b) increase of settlements from embankment construction until last measurement (Alexiew and Raithel, 2015)

GEC-supported and -unsupported zones are also shown for a better understanding. According to the measured data, the following results were obtained.

The settlements of the unsupported shoulder were – despite the lower average load – about 3 times higher than the GEC-supported full-height part of embankment.

The consolidation process was quick, especially in the GEC-supported zone: after only two months of embankment construction, practically no further increase was found.

The total settlement in the GEC zone showed a maximum of around 10 cm, which is a very modest value.

4.3.5 Bastions Vijfwal Houten, Netherlands, 2005

The area of Houten-Zuid was one of the locations, which the Dutch government had targeted as a growing area for housing development in the Netherlands. In two of these areas in Houten-Zuid, landscape embankments (so-called Bastions West and East) were planned as a connection between the residential area and the surrounding natural landscape. Due to the soft subsoil, settlements of 1.6–1.9 m for Bastion West and 0.5–0.8 m for Bastion East were expected to occur. The problems faced were not only the settlements as such, but also the long consolidation time and the possible lateral pressure in depth on the rigid pile foundations of the already constructed adjacent buildings. After checking different options, the foundation of the "Bastions" on GECs was chosen as the optimal one. In Figure 4.22, the plan view and a typical cross-section of Bastion West are shown, as are the main parameters of the project listed in Table 4.5.

To ensure the appropriate compaction of the column fill material and the estimated load-bearing capacity of the columns, penetration tests inside the columns and simple short-term load tests were carried out as shown in Figure 4.23a. Figure 4.23b shows typical results of the settlement measured on the top of the GECs and between them with a final settlement of about 0.35 m completed in 180 days. It is observed that the calculated settlement met very well the measurement, particularly with that between the GECs.

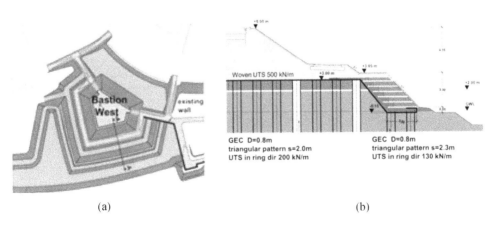

(a)   (b)

Figure 4.22 (a) Plan view of Bastion West; (b) a typical cross-section with GECs (Alexiew and Raithel, 2015)

Table 4.5 Main data of the project Bastions at Houten (Alexiew and Raithel, 2015)

Embankment:	Bastion West	Bastion East
height	5.5 m	5.5 m
fill material	γ = 17 kN/m³ / φ' = 20° / c´ = 2 kN/m²	γ = 17 kN/m³ / φ' = 20° / c´= 2 kN/m²
traffic load	20 kN/m²	20 kN/m²
Soft Soil Layer:	organic clay & peat	sandy organic clay
thickness	7.5 m	3.0 m
properties	γ = 14 kN/m³ / φ' = 17° / c´ = 2.5 kN/m² $E_{s.pref}$ = 2000 kN/m² (p.ref = 100 kN/m²)	γ = 17 kN/m³ / φ' = 22.5° / c´= 2 kN/m² $E_{s.ref}$ = 3000 kN/m² (p.ref = 100 kN/m²)
ground water level	- 2.0 m	n.a.
Foundation System:	geosynthetic encased columns	geosynthetic encased columns
geometry	s = 2.00 m dc = 0.80 m	s = 2.30 m dc = 0.80 m
column fill	γ = 19 kN/m³ / φ' = 32.5° / c = 0 kN/m² (sand)	γ = 19 kN/m³ / φ' = 32.5° / c = 0 kN/m² (sand)
encasement	Ringtrac® 3500 PM UTS = 200 kN/m J_K = 3500 kN/m J_d = 2100 kN/m	Ringtrac® 2000 PM UTS = 130 kN/m J_K = 2000 kN/m J_d = 1000 kN/m
basal reinforcement	Stabilenka® 500/100 UTS = 500 kN/m	Stabilenka® 500/100 UTS = 500 kN/m
estimated settlements	≤ 0.40 m	≤ 0.15 m

* J_K = short term radial tensile stiffness; J_d = long term radial tensile stiffness (120 years)

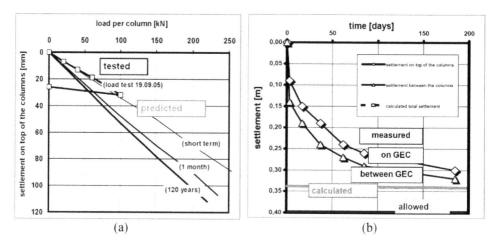

Figure 4.23 (a) Short-term load test on a single GEC; (b) long-term settlements on top and between GECs (Alexiew and Raithel, 2015)

4.3.6 Steel plant TKCSA, Itaguaí, Brazil, 2006–2010

The steel plant had to be built on extremely soft soil lowlands on the seashore inclusive of a large, heavy loaded stock yard for coal, coke, ore, and additives (Alexiew *et al.*, 2010). Figure 4.24 shows a typical soil profile down to 18 m below terrain with an undrained strength varying from 5 to 10 kPa. A wide range of factors had to be taken into consideration regarding the foundation of the very large railroads (Alexiew *et al.*, 2010). The final decision was to found them on GECs in the coal/coke area. A total of about 250 km of GECs were installed with a diameter of 78 cm at a center-to-center spacing of 2 m, square mesh pattern, average length of 10 m, and varying strength.

In an early stage of stockyard construction, an instrumented load test was performed to study the behavior of a full stock bed and an adjacent GEC-supported runway, both on high-strength, low-strain horizontal geosynthetic reinforcement with ultimate tensile force up to 1600 kN m^{-1}. Figure 4.25 shows some test results. Note that there are no GECs installed to the left.

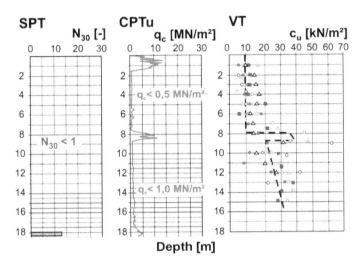

Figure 4.24 Steel plant TKCSA: typical geotechnical profile (Alexiew and Raithel, 2015)

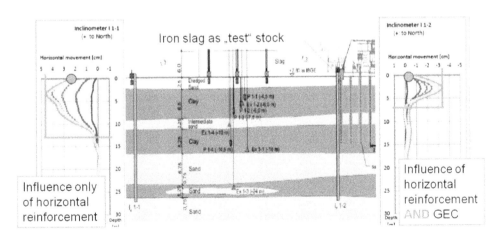

Figure 4.25 Different horizontal displacement behavior without (left) and with (right) adjacent GECs (Alexiew and Raithel, 2015)

The following are some of the most important results:

- The consolidated settlements due to runway platform of 2–3 m thickness, ballast bed, sleepers, etc., amounted to 20 cm after two months;
- A 750-ton stacker/reclaimer kept in a fixed position for a week provoked 4 cm of additional settlement; and
- A similar test with rotated boom (for example, under maximal load eccentricity) resulted in a cross-tilting of runway of < 0.25%.

This data met the requirements concerning the short- and long-term deformations; thus, these tests gave the final "green light" for the concept of a "runway on GEC."

4.4 Final remarks

The engineering applications of GEC described in this chapter have included stockyards, railroads, airports, dykes, and housing developments more familiar to the authors. However, GEC has been used in many more projects around the world, e.g., in Spain, Malaysia, etc., but the authors know no publications related to those projects. However, it is worth mentioning that the application in Malaysia (unpublished) was quite specific: a row of GECs filled with a mixture of sand and bentonite was installed with zero space between them as a curtain to reduce (to 90%, but not stop completely) ground water stream – it worked. Note that even the case studies overview in Alexiew and Raithel (2015) does not include all the projects completed.

Chapter 5

Application of numerical analyses

5.1 Introduction

By the beginning of the 2000s, numerical analyses using either finite element or finite difference methods have been frequently employed to evaluate the behavior of the geosynthetic encased column (GEC) system. They can simulate the interaction mechanisms between soil and geosynthetic material by adopting the stress-strain coupled formulation with reasonable accuracy. Numerical analyses, especially using the finite element method, provide a more fundamental understanding of GEC behavior through parametric studies to investigate the influence of the input parameters, which were mostly verified with experimental investigations. Several two- and three-dimensional finite element analyses were performed to study the influence of critical parameters, such as area replacement ratio, stiffness modulus of the encasement material, thickness of clay layer, embankment stress, and reinforcing modes. Some of the most important numerical analyses of the GEC system are listed in Table 5.1.

This chapter presents the results of the numerical analysis of the TKCSA test embankment described in Chapter 4. Axisymmetric and plane strain simulations are performed using Plaxis 2D and 3D finite element codes (Brinkgreve and Vermeer, 2012), and the results are compared with field measurements. The 2D axisymmetric analysis is carried out by modeling a singular encased granular column localized below the embankment centerline accompanied by its surrounding soft soil (i.e., unit cell concept). 2D plane strain analysis is conducted by simulating a section crossing the centerline of the embankment using Raithel and Henne's (2000) method to convert the encased column into an equivalent wall. In addition, a full strip 3D analysis is modeled to study the influence of the basal geogrid reinforcement. The model configuration, calculation steps, and results of the numerical analyses are presented with details in this chapter, thus the applicability and limitations of each model are discussed.

5.2 2D axisymmetric analysis

The unit cell concept is applied to perform the axisymmetric analysis to determine the settlements and the stresses separately acting on the encased column and the surrounding soft soil, as well as the pore pressures and geotextile hoop strains. The columns localized in the central area of the test embankment are arranged with an average center-to-center spacing (s) of 2.0 m along a square grid, resulting in a 2.26 m diameter unit cell ($d_e = 1.13s$). The model configuration, boundary conditions, and finite element mesh are shown in Figure 5.1. The

Table 5.1 Some important numerical analyses of GEC system

Reference	Analysis type	Case considered	Parameters analyzed
Murugesan and Rajagopal (2006)	2D unit cell	Hypothetical model	d_c, J, H_{em}
Yoo and Kim (2009)	2D unit cell and 3D analyses	Actual model	Comparison of 2D and 3D analyses
Khabbazian et al. (2010)	3D unit cell	Hypothetical model	d_c, J, ϕ_c, L_{enc}, L_{col}
Yoo (2010)	2D unit cell and 3D analyses	Actual model	J, a_E, L_{enc}, H_{em}
Tandel et al. (2012)	3D analysis	Hypothetical model	J, E_s, d_c, E_c
Keykhosropur et al. (2012)	3D analysis	Actual model	ϕ_c, E_c, d_c, J
Almeida et al. (2013)	2D unit cell	Hypothetical model	J, H_{em}, H_s
Mohapatra et al. (2017)	3D unit cell	Prototype scale	σ_3, d_c, mesh pattern
Khabbazian et al. (2015)	3D unit cell, 2D unit cell and 3D strip	Hypothetical model	σ_v, δ_h, δ_v

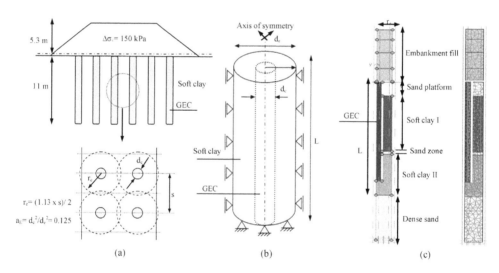

Figure 5.1 Axisymmetric analysis of TKCSA test embankment: (a) geometrical data of the test embankment and encased columns; (b) unit cell concept; (c) model adopted and finite element mesh (Hosseinpour et al., 2015)

finite element mesh adopted was based on the mesh sensitivity analysis. A fine grain mesh was assigned for the whole model with a local mesh refinement close to the encased stone column.

The vertical and horizontal displacements are restrained at the bottom boundaries, but the model is free for the vertical displacements at the lateral borders.

The geotextile casing and the basal geogrid are simulated using a geogrid element, a slender structure that only sustains an axial tensile force along its length, with stiffness values equal to 1750 and 2200 kN m^{-1}, respectively. As far as the constitutive models are

Table 5.2 Material parameters used in finite element analysis of the test embankment (Hosseinpour et al., 2015)

Material and constitutive model	γ_{sat} $(kN\,m^{-3})$	k_h $(m\,d^{-1})$	k_v $(m\,d^{-1})$	ϕ' $(°)$	C' (kPa)	E' (kPa)	C_c $(-)$	C_s $(-)$	C_k $(-)$
Embankment fill (MC)	28	1	1	38	0	50000	–	–	–
Granular column (MC)	20	10	10	40	0	80000	–	–	–
Soft clay I (SS)	14.4	9.7×10^{-6}	6.2×10^{-6}	25	2	–	0.98	0.084	1.03
Soft clay II (SS)	17.2	8.8×10^{-6}	5.6×10^{-6}	28	3	–	0.13	0.040	1.03
Working platform (MC)	19.5	0.6	0.6	33	0	15000	–	–	–
Dense sand (MC)	20	1	1	38	0	30000	–	–	–
Sand lens (MC)	18.5	0.5	0.5	30	0	22000	–	–	–

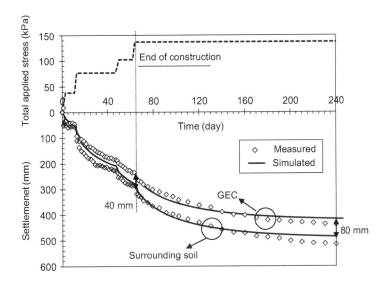

Figure 5.2 Predicted and measured settlements on encased column and surrounding soil (construction and post-construction period) (Hosseinpour et al., 2015)

concerned, the elastic perfectly plastic model with Mohr-Coulomb (MC) failure criteria was adopted for both the encased column and the embankment material. Soft clay behavior was simulated using Soft Soil (SS) elasto-plastic model and the parameters were obtained from the site investigation (Hosseinpour *et al.*, 2017) as shown in Table 5.2.

5.2.1 Settlement estimation

Measured and calculated settlements (through finite element analyses) on the top of the GEC and the surrounding soft soil are shown in Figure 5.2. It is observed that the 2D axisymmetric analysis predicted the measured settlements reasonably well, in particular during construction and to some extent for the stage consolidation intervals. Both measured and simulated results showed that the settlement increased notably just after each layer was placed and also during consolidation when excess pore pressure was dissipated.

Unlike the analytical calculations (Raithel and Kempfert, 2000; Castro and Sagaseta, 2011), a difference was seen between settlements at the top of the GEC and at the surrounding soil. Similar to the total settlement, this difference increased with time as a differential settlement of around 80 mm occurred at the end of monitoring time for both measured and computed values. This differential settlement is caused by the different stiffness values of the GEC and the soft clay (Almeida *et al.*, 2013).

5.2.2 Total vertical stresses

The total vertical stresses measured (with total pressure cells) on the top of the GEC and the surrounding soil (i.e., midpoint between the columns) are compared with the values obtained from finite element analysis (FEA) in Figure 5.3. It is observed that the measured stress distribution is simulated reasonably well by the axisymmetric analysis. The results also show that the total vertical stresses acting on the GEC tend to increase continuously during post-construction. Inversely, the total vertical stress acting on surrounding soil (midpoints between the columns) decreased at a slow rate. This behavior can be attributed to the decrease in the apparent stiffness of surrounding soil from quasi-undrained stiffness to drained stiffness during consolidation. Comparison of field data and the results of finite element analysis show that 2D axisymmetric analysis gave a satisfactory estimation of the total vertical stresses, particularly at midpoints between the columns.

Measured and predicted stress concentration ratios ($n = \Delta\sigma_{v,c}/\Delta\sigma_{v,s}$) are shown in Figure 5.4. It is seen that both the measured and the computed stress concentration increased during construction after each layer was placed, as well as during the post-consolidation period

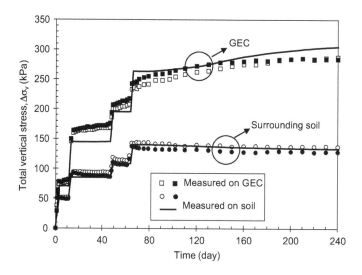

Figure 5.3 Vertical stresses acting on GEC and surrounding soil: measured and predicted results (Hosseinpour *et al.*, 2015)

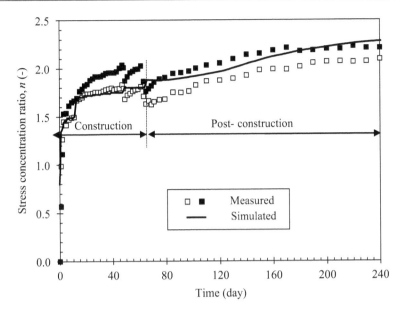

Figure 5.4 Measured and simulated stress concentration ratio (n) (Hosseinpour *et al.*, 2015)

due to change in apparent undrained stiffness of the soft clay. The increase in stress concentration during post-construction can also be attributed to the development of soil arching resulting from further stretching of the geotextile encasement, which caused the column to support more total vertical stress. Since greater total vertical stress was transmitted to the encased column, stress concentration increased almost continuously.

5.2.3 Excess pore pressures

The variations of the excess pore pressure with elapsed time for the piezometers located at 3 m (PZ1), 6 m (PZ2), and 8 m (PZ3) below the ground surface and on the embankment centerline are shown in Figure 5.5. A sharper increase can be observed just after the placing of each layer, and subsequently the excess pore pressure dissipates partially during consolidation intervals. It can be seen that 2D axisymmetric analysis simulated both excess pore pressure buildup and dissipation for the stage construction of the embankment reasonably well, but there is almost 20% difference in the maximum excess pore pressure in loading stage four, which could be related to the actual thickness of the layer placed in this loading stage. The faster dissipation of excess pore pressure obtained from FEA can be explained by the actual columns layout in the test area. The GEC were installed in an irregular square pattern with center-to-center spacing ranging between 1.75 and 2.25 m. The analysis, however, was performed using a unit cell generated considering 2.0 m center-to-center spacing between the columns, which is the average value obtained in the central area of the test embankment, relevant for the axisymmetric analysis. Thus the localized larger spacing between the columns, where piezometers are installed, could be the reason for slower dissipation of pore pressure obtained by measurement compared with the FEA.

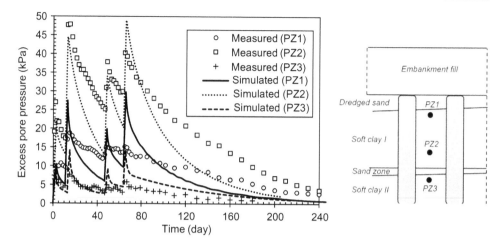

Figure 5.5 Measured and simulated excess pore pressures (Hosseinpour et al., 2015)

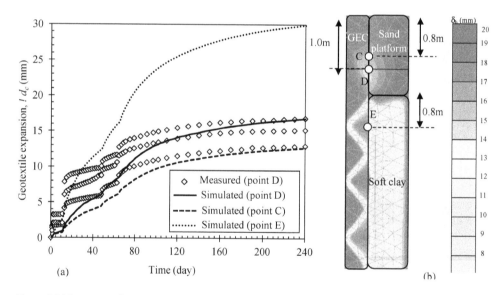

Figure 5.6 Variations of the horizontal deformation: (a) measured and computed geotextile expansion; (b) distribution of the horizontal deformation (Hosseinpour et al., 2015)

5.2.4 Geotextile expansion (column bulging)

Variations of the measured and simulated geotextile expansion ($[\Delta d_c]$) are shown in Figure 5.6a. The geotextile expansion (i.e., column radial deformation) increased notably just after each layer was placed and increased continuously during post-construction. Continuous encasement expansion coincides with the variation of the settlement and the total vertical stress measured on the GEC. The results of the 2D axisymmetric finite element analysis are used to compare the geotextile expansion at 0.8 m below the soft clay (i.e., point E in Fig. 5.6b), equal to the column diameter, with the field data measured at point D. It is observed that the simulated column horizontal deformation at point E was about twice the value recorded by instrumentation (i.e., point D). The smaller horizontal deformation measured at point C (i.e., 0.8 m below the ground surface), compared with point E, can be attributed to the contribution of the sand working platform, which provides higher confining support acting on the column along this zone.

5.2.5 Influence of the column spacing

The 2D axisymmetric analysis was applied to study the influence of the spacing between the columns, using s = 1.75 and 2.25 m as the minimum and the maximum spacing between the GEC in the test area respectively. Figure 5.7 compares the settlement measured at midpoint between GEC with those obtained from the finite element analysis. The larger the spacing between the GEC, the greater settlement observed. For instance, the settlement computed assuming s = 2.25 m was around 23% greater than the settlement computed assuming s = 1.75 m. It was also observed that the settlement computed assuming s = 2.0 m matches with the measured data, particularly at the end of monitoring time, which confirms that the columns spacing of s = 2.0 m used for the unit cell localized in the central area of the test embankment is a representative value.

In Figure 5.8, variations of the measured total vertical stress on the top of the GEC are compared with the simulated results, and it was observed that for a constant column diameter, the larger spacing between the columns (a lower the area replacement ratio) caused the column to support higher total vertical stress. For instance, at the end of monitoring time, the total vertical stress on the top of GEC spaced 2.25 m was approximately 20% greater than that for s = 1.75 m.

In Figure 5.9, variations of the measured total vertical stress acting on the top of the surrounding soft soil are compared with the numerical results. Unlike vertical stress acting on the GEC, the total vertical stress acting on surrounding soil reduced as spacing between the columns increased (i.e., the area replacement ratio was lower). As clearly seen, the best

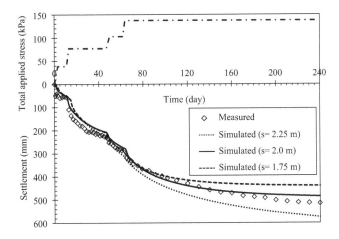

Figure 5.7 Influence of the spacing between the columns on settlements on the top of the surrounding soil

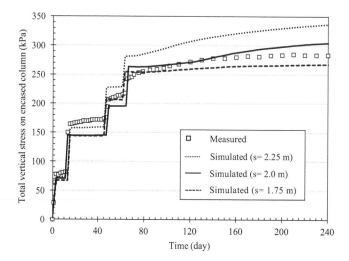

Figure 5.8 Influence of the spacing between the columns on variation of the total vertical stress acting on encased column

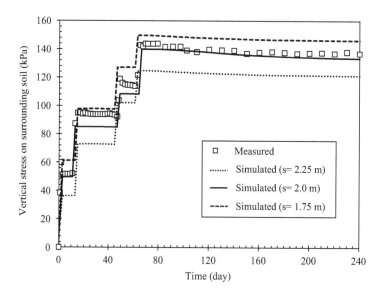

Figure 5.9 Influence of the spacing between the columns on variations of the total vertical stress on the top of the surrounding soil

prediction of the measurements was achieved when finite element analysis was performed assuming s = 2.0 m, which is the average value at the central area of the test embankment. Comparison of Figures 5.8 and 5.9 indicates that the GEC spaced larger improved the stress concentration ratio, as also reported by experimental investigations (Murugesan and Rajagopal, 2007) and some numerical studies (Khabbazian et al., 2010; Hosseinpour et al., 2014). For example, the stress concentration ratio (n) at the end of monitoring time is equal to 3.1 by assuming s = 2.25 m; however, the stress concentration ratio reduces to 1.86 when s = 1.75 m.

5.3 2D plane strain analysis

The typical 2D axisymmetric unit cell model with horizontally fixed side boundary conditions is not suitable to simulate horizontal displacements due to its boundary conditions. Although a 3D analysis is more appropriate for this type of geosynthetic application, a 2D plane strain model might be also used to predict the horizontal deformation of the foundation soil underneath the embankment. A simplified methodology for a 2D plane strain analysis of unencased granular columns was proposed by Tan *et al.* (2008), in which a unit cell of a granular column was modeled into a wall in order to obtain the equivalent plane strain column width. The significance of the geotextile stiffness based on a 2D plane strain model is ignored, resulting in computed total deformations higher than the actual values. Therefore, the effect of the geotextile encasement is indirectly taken into account by an increased frictional angle for the column material introduced by Raithel and Henne (2000), as discussed in Chapter 2.

5.3.1 Model configuration

As shown in Figure 5.10, the axisymmetric to plane strain transformation was performed according to the method (see Chapter 2) proposed by Tan *et al.* (2008), which was also successfully used by Almeida *et al.* (2014). In this methodology, the granular column is

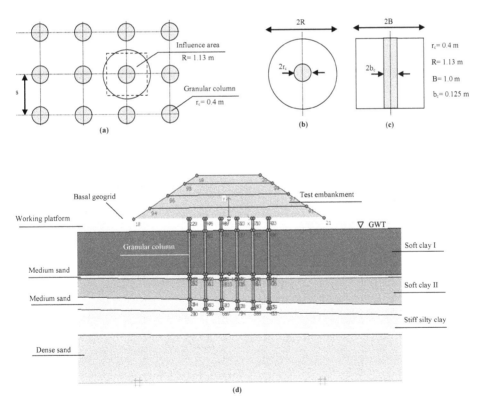

Figure 5.10 Plane strain configuration of the test embankment: (a) layout of granular columns; (b) axisymmetric unit cell; (c) converted plane strain model; (d) numerical model adopted for finite element analysis (Hosseinpour *et al.*, 2017)

substituted by an equivalent plane strain wall, in which the half of column width b_c is determined by Equation (2.29).

Given geometrical data of the granular columns yields to an equivalent plane strain column width (i.e., $2b_c$) equal to 0.25 m; however, the geotextile encasement effect is not yet taken into account. The influence of the geotextile encasement was applied by a change in the friction angle of the column proposed by Raithel and Henne (2000) using Equation (2.35). This formulation is derived using Mohr's circle of stresses and by the stress difference ($\Delta\sigma_{3,geo}$) between the whole horizontal supporting stress from the soft soil ($\sigma_{h,s,total}$) and the horizontal inner stress from the granular column ($\sigma_{3,c}$). It is noted that the values of stress difference, horizontal stresses of soft soil, and horizontal stresses inside the column are not constant along the column's depth at any stage of embankment construction, determined by axisymmetric analysis.

Therefore, the equivalent friction angle was calculated based on the average values of those variables over the entire depth of column and the average values over all stages of loading. Using Equation (2.35) and the average values of the horizontal stresses, computed by the axisymmetric analysis, yields to an equivalent friction angle (ϕ'_{sub}) equal to 66° for granular column material to be used in the plane strain analysis. The material properties and constitutive models are the same as those used in axisymmetric analysis presented in Table 5.2.

5.3.2 Settlement estimation

According to Figure 5.11, the settlements computed using the equivalent friction angle approach (i.e., with encasement effect) predict the measured values fairly well, particularly for that measured on the top of the surrounding soil. However, the settlement computed using the column original friction angle (i.e., without encasement effect) was greater than the measured value. The influence of the geotextile encasement is clearly observed in the plane strain results, as the settlement using the column original friction angle was about twice of

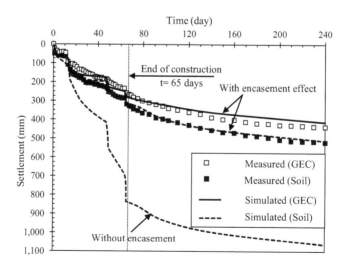

Figure 5.11 Measured and predicted (FEA) settlements vs. time with and without encasement effect (Hosseinpour et al., 2017)

that computed using geosynthetic-equivalent friction angle approach. This is because the equivalent frictional angle would cause the column to have more shear strength, resulting in a smaller column radial deformation, subsequently in less settlement under the equal applied load. In general, the geosynthetic-equivalent friction angle used in the plane strain analysis appears to be a suitable approach for settlement estimation of the test embankment over GEC for the section crossing the embankment centreline.

5.3.3 Soil horizontal displacement

Horizontal displacements beneath the test embankment toes measured by inclinometers, with their position shown in Figure 4.2, are compared to those obtained by 2D plane strain analysis in Figure 5.12. It is observed that the horizontal deformations obtained using the equivalent fraction angle approach (i.e., with encasement effect) showed reasonable agreement with measurements beneath the depth of 3.0 m, particularly with the magnitude of the maximum horizontal deformation. However, the maximum soil horizontal deformation computed using the column original friction angle (i.e., without encasement effect) was as much as 3–5 times greater than the values measured at the end of construction and end of monitoring time, respectively.

Figure 5.13 illustrates the distribution of the horizontal deformation in soil layer with and without the encasement effect. A larger horizontal deformation is observed while the geotextile confining support is ignored. Concentration of the large horizontal deformations

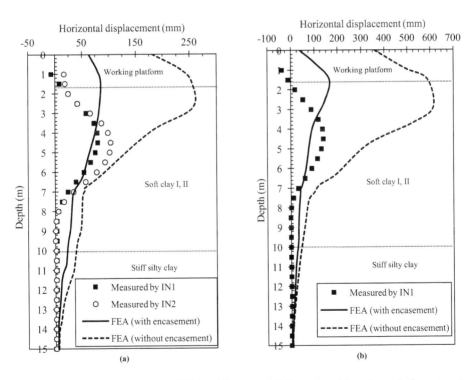

Figure 5.12 Measured and predicted (FEA) soil horizontal deformation: (a) at the end of construction (i.e., 65 days); (b) at the end of monitoring period (i.e., 240 days) (Hosseinpour *et al.*, 2017)

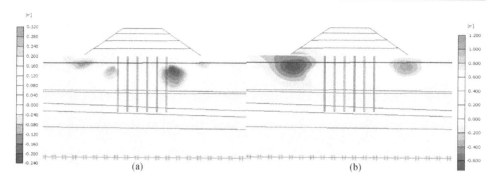

Figure 5.13 Distribution of the incremental horizontal deformation at the end of monitoring time: (a) using equivalent friction angle; (b) using original friction angle (Hosseinpour *et al.*, 2017)

beneath the embankment toes, shown in Figure 5.13b, provides clear evidence of embankment lateral spreading taking place when geotextile encasement is not taken into account.

Generally, it can be said that the use of the geosynthetic-equivalent friction angle (i.e., contribution of the geotextile encasement) resulted in a much more proper computation of the soil horizontal deformation occurring beneath the embankment toes. However, the computed values were overpredicted in the upper 3.0 m. The simple Mohr-Coulomb model adopted for the granular column may be the reason behind the difference between measured and predicted horizontal deformation in this zone. In any case, it is relatively common to obtain overall good agreement between observations and numerical predictions in terms of vertical deformation and pore pressure, but not for the horizontal displacement (Almeida *et al.*, 1986).

5.3.4 Excess pore pressures

Variations of the excess pore pressures measured by piezometers PZ1 and PZ2, both installed below the embankment centerline but in different depths (PZ1 at −8 m depth and PZ2 at −6 m depth), are compared to the predicted values in Figure 5.14. It can be seen that the geosynthetic-equivalent friction angle approach (i.e., with encasement effect) predicted reasonably well the pore pressure buildup and some extent of its dissipation. Using the column original friction angle (i.e., without encasement effect), however, resulted in a greater maximum excess pore pressure and a longer dissipation time compared to the measurement, which could be related to the higher total vertical stress transmitted to the surrounding soil when the influence of the encasement is ignored.

5.3.5 Total vertical stresses

The distribution of the total vertical stress below the embankment at the end of monitoring time is demonstrated in Figure 5.15. It is seen that the use of the geosynthetic-equivalent friction angle approach caused the total vertical stress in soft clay to decrease and, inversely, the total vertical load supported by the encased columns to increase. In other words, contribution of the geotextile encasement enhanced the load transfer mechanism through the soil arching over the granular columns. This behavior clarifies that the simulation of the

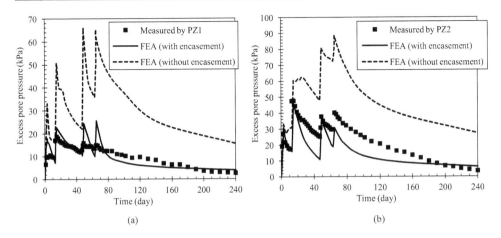

Figure 5.14 Variations of the measured and predicted (FEA) excess pore pressures with and without encasement: (a) measured by PZ1; (b) measured by PZ2 (Hosseinpour *et al.*, 2017)

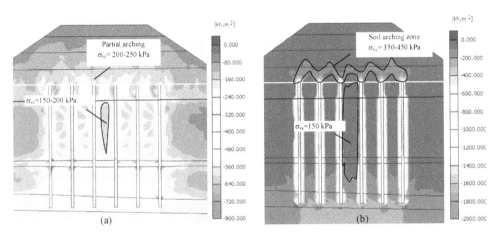

Figure 5.15 Distribution of the total vertical stress at the end of monitoring period: (a) using column original friction angle; (b) using geosynthetic-equivalent friction angle (Hosseinpour *et al.*, 2017)

geotextile encasement effect, using the equivalent friction angle approach, predicted the variations of the excess pore pressures reasonably well at the same depth that the piezometers were installed.

5.3.6 Stability analysis of embankment

In order to verify the stable performance of the test embankment, c-phi reduction stability analysis (Brinkgreve and Vermeer, 2012) was performed to evaluate the factor of safety of the test embankment at two consecutive stages: just after load application, and just before next load increment. The stability analysis was conducted using the original column friction angle and the equivalent friction angle to assess the encasement influence on the factor of safety of the test embankment.

Results of the stability analysis of the test embankment are summarized in Figure 5.16 in plots of the factor of safety (FS) against the height of embankment. Regardless of the geotextile influence, it is seen that the minimum factor of safety is reached just after the load application when the excess pore pressure in clayey foundation is maximum (see Fig. 5.17a). Upward trends for the constant height of embankment denote an increase in the factor of safety during the consolidation interval following the loading stage. This increase is because of the increase in shear strength of the soft clay due to dissipation of excess pore pressure, as illustrated in Figure 5.17b. As expected, the largest improvement is seen for loading stage 4 with about six months consolidation period, followed by loading stage 2, which had the longest consolidation interval (i.e., 35 days) compared to loading stages 1 and 3. Concerning the geotextile encasement influence, a significant improvement in factor of safety is reached as geosynthetic-equivalent friction angle was used. It is also seen that when the geotextile effect was ignored, the test embankment failed in loading stage 3 due to insufficient shear

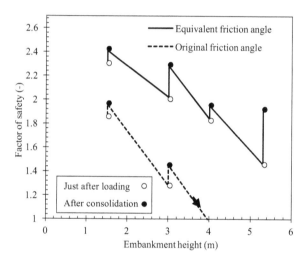

Figure 5.16 Predicted factor of safety of the test embankment during construction and consolidation stages (Hosseinpour *et al.*, 2017)

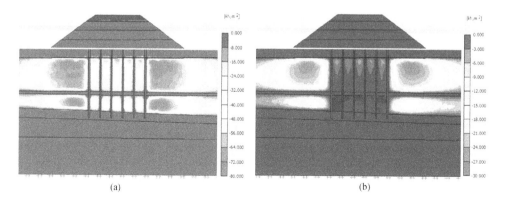

Figure 5.17 Distribution of excess pore pressure: (a) end of construction; (b) end of monitoring period (Hosseinpour *et al.*, 2017)

strength of the soft foundation. As failure had not occurred in the field, the use of original friction angle of column material (not considering the contribution of geotextile) is not representative of the field condition.

5.4 3D strip analysis

Three-dimensional 3D numerical analysis was applied in order to verify the contribution of the working platform and the basal geogrid reinforcement on settlement and stress distribution below the embankment. Accordingly, a rectangular slice located below the embankment centerline, where the instrumentation was concentrated, was selected to perform the numerical simulation. The zone of interest has two orthogonal planes of symmetry, thus only half of the test embankment over reinforced ground was modeled. The slice adopted is 1 m wide including three central encased granular columns, as illustrated in the plan and cross-sectional views in Figure 5.18.

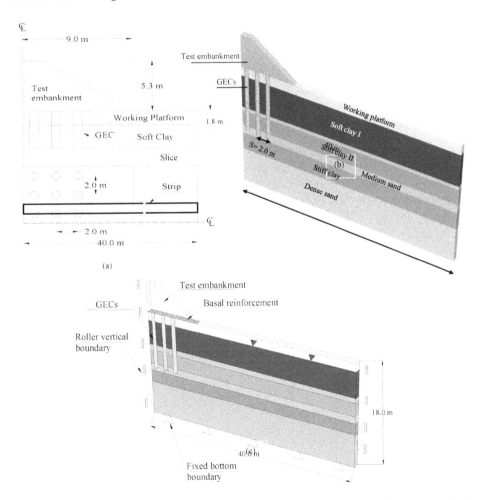

Figure 5.18 3D configuration of the embankment over GECs: (a) slice considered for numerical analysis; (b) finite element model; (c) boundary conditions

The basal geogrid reinforcement and the geotextile encasement were modeled as isotropic nonlinear geogrid elements, available in PLAXIS 3D, which are composed of six node triangular surface elements with three translational degrees of freedom per node. Geogrid elements have only axial stiffness, no bending stiffness; therefore, they can only sustain tensile forces along their length. A perfect interface adherence (i.e., no interface element) was assigned between the geogrid element and the adjacent soil. Several studies showed that the assumption of the perfect interface adherence in working stress conditions results in reasonable predictions with respect to measured data (Hatami and Bathurst, 2005; Tandel et al., 2012).

The lateral extent of the model was then chosen to be 40 m to avoid any influence of the outer boundary. Considering the boundary fixities, the model was restricted to deform horizontally on the vertical sides (i.e., roller boundaries) while fully fixed along the base, as seen in Figure 5.18. The groundwater table level was set to the interface between the working platform and the top soft clay layer, as observed *in situ*.

The load application was simulated by the current stage of construction of the test embankment followed by a period of six months to allow for consolidation. The steps of the calculation consisted of activating the clusters corresponding to the various embankment layers in order to simulate embankment construction, followed by consolidation intervals between the stages to analyze the development and dissipation of the excess pore pressures in the saturated soft soil as a function of time.

5.4.1 Settlements

Figure 5.19 shows the influence of the working platform thickness on the settlement at the mid-point between encased columns (point A) calculated by 3D numerical modeling. Three different values of the working platform thickness were analyzed ($H_{wp} = 0.0$ m; 0.5 m; and 1.5 m) while the encasement stiffness and basal geogrid stiffness were kept constant as $J_{en} = 1750$ kN m^{-1} and $J_{bs} = 2200$ kN m^{-1}, respectively. Although embankment construction on very

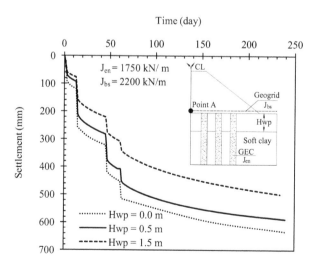

Figure 5.19 Influence of working platform on settlement computed at point A

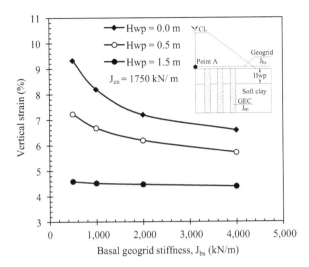

Figure 5.20 Combined influence of basal geogrid and working platform on maximum settlement (point A)

soft clay layers without a working platform is practically impossible, analyses were also conducted without the working platform to assess its influence on the results. An increase of the working platform thickness causes the settlement below the embankment centerline to be reduced in either the construction stages or the post-construction period.

Figure 5.20 shows the vertical strain as defined by the maximum settlements at point A normalized by the thickness of the working platform and soft clay for different basal geogrid stiffness. These values were taken at the end of the monitoring time (i.e., 240 days) when around 90% of consolidation was completed. It can be seen that an increase in basal geogrid stiffness reduces the vertical strain; nevertheless, its influence is associated with the working platform thickness. For example, without a working platform the vertical strain for upper and lower limits of geogrid stiffness (i.e., J_{bs} = 500 and 4000 kN m^{-1}) differ by about 2.5%. However, when the working platform thickness is equal to 1.5 m, an increase in geogrid stiffness does not significantly affect the vertical strain values, as reported by previous researchers (King *et al.*, 1993; Han and Gabr, 2002; Zhang *et al.*, 2015). According-ing to the results, the influence of the basal geogrid stiffness on the maximum vertical strain is closely associated with the thickness of the granular bed placed below the embankment.

5.4.2 Total vertical stresses

Figure 5.21 shows the influence of the working platform (H_{wp}) on the stress concentration factor (SCF), defined by the average vertical stress on the encased column (point B) to the average vertical stress on the surrounding soil (point A). In these analyses, the encasement stiffness was 1750 kN m^{-1} and the SCF was calculated just after each load application stage.

It is observed that the SCF is roughly constant for H_{wp} = 1.5 m, fairly close to the mea-sured value of 2.3 during construction stages and post-construction (Almeida *et al.*, 2014). As seen, the influence of the geogrid stiffness on the SCF is quite significant in the absence of a working platform as the stiffer geogrid reduces the SCF values. Nevertheless, the SCF is

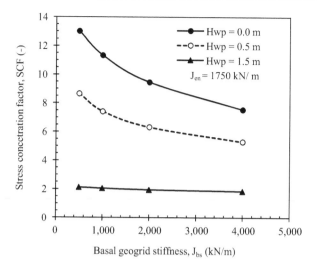

Figure 5.21 Combined influence of basal geogrid and working platform on SCF

almost constant for any basal geogrid stiffness when a 1.5-m thick working platform exists. In other words, there is an optimum thickness for a working platform (i.e., 1.5 m for the present case), since beyond that value an increase in geogrid stiffness would not enhance the soil arching below the embankment.

5.4.3 Soil horizontal displacement

The influence of the working platform on the maximum horizontal displacements, which typically occurred at the clay-working platform interface, are shown in Figure 5.22a. It can be seen that the working platform reduced the maximum soil horizontal displacement at all stages of embankment construction.

Figure 5.22b shows the maximum horizontal displacement as a function of the basal geogrid stiffness. For the case analyzed without and with a 0.5 m thick working platform, an increase in basal geogrid stiffness reduced the maximum soil horizontal displacement, thus also improving the global stability of the test embankment. However, similarly to the previous comparisons, when the working platform thickness was 1.5 m, a change in the geogrid stiffness did not affect the maximum horizontal displacement. In this case, the low contribution of the basal reinforcement is reflected in the low values of the mobilized tensile forces and in the slight variation of the maximum soil horizontal displacement with increasing reinforcement stiffness.

5.4.4 Tensile force in geogrid reinforcement

The maximum tensile forces developed on the basal geogrid (T_{max}) are plotted in Figure 5.23 as a function of the embankment height. It can be observed that the maximum tensile force decreases as the thickness of the working platform increases, from $T_{max} = 164$ kN m^{-1} for $H_{wp} = 0.0$ m to $T_{max} = 20$ kN m^{-1} for $H_{wp} = 1.5$ m, both calculated at the end of construction. It is interesting to note that when the working platform does not exist ($H_{wp} = 0.0$ m), the

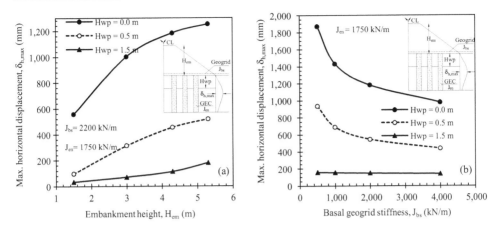

Figure 5.22 Variation of maximum soil horizontal displacement: (a) influence of working platform; (b) influence of basal geogrid stiffness

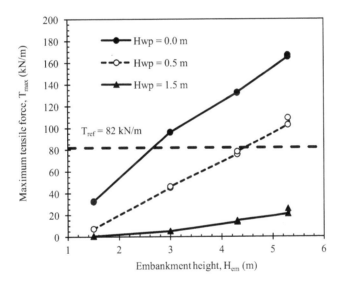

Figure 5.23 Maximum tensile force in basal geogrid for different working platform thickness

mobilized tensile force at the geogrid reinforcement crosses the allowable tensile force at the second stage of embankment construction. But when a 1.5 m thick working platform is used ($H_{wp} = 1.5$ m), the maximum mobilized tensile force is about 3 times lower than the allowable value.

Figure 5.24 compares profiles of the tensile forces mobilized along the basal geogrid at the end of monitoring time as the encasement stiffness changes. These analyses were carried out with the basal reinforcement stiffness and working platform thickness equal to $J_{bs} = 2200$ kN m^{-1} and $H_{wp} = 1.5$ m, respectively. It can be observed that an increase in encasement stiffness reduces the tensile force along the basal reinforcement. As reported by previous studies

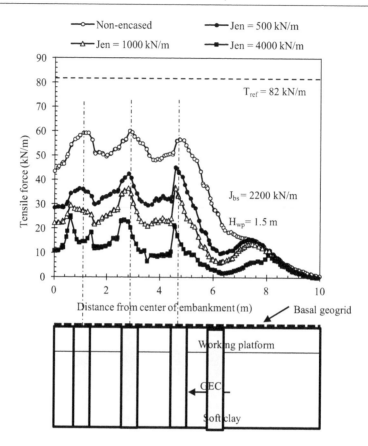

Figure 5.24 Influence of encasement stiffness on tensile force along the basal geogrid

(Murugesan and Rajagopal, 2006; Gniel and Bouazza, 2010), the stiff geosynthetic encasement enhances the part of the load transferred directly to the top of the encased columns and, subsequently, the basal reinforcement is less responsible for carrying the embankment applied stress.

It can also be observed that for two limit cases of encasement stiffness, i.e., J_{en} = 4000 kN m^{-1} and J_{en} = 500 kN m^{-1}, the maximum tensile force along the basal geogrid is about 28% and 55% of the allowable tensile force, respectively (T_{ref} = 82 kN m^{-1}). This ratio, however, increases up to 73%, as columns are not encased. This is indirect evidence of the effectiveness of the encasement in controlling the soil horizontal deformations.

5.5 Final remarks

Each type of analysis has its advantages, disadvantages, and limitations. Table 5.3 compares the capabilities of some numerical and analytical models most usually performed with respect to conditions and variables of main interest. The 2D unit cell is the most common numerical analysis performed, and it provides the main necessary information for a typical geotechnical design in general. The 3D unit cell analysis is more rigorous than the 2D unit cell analysis, but the differences obtained may be relatively small, thus it may not be

Table 5.3 Capabilities of numerical and analytical models

Conditions/Variables	2D unit cell (or 3D unit cell)	3D strip	2D Plane strain	Raithel and Kempfert's (2000) Analytical model
Heterogeneous soil conditions	***	***	***	
Central settlement at embankment base	***	***	**	
Differential settlement at the top of the embankment	**	***	**	–
Settlements along the embankment base	–	***	**	–
Geosynthetic tensile force at the GEC	***	***	–	**
Vertical stresses on column and soil	**	***	*	**
Horizontal displacements at embankment base	–	***	*	–
Geosynthetic tensile force at the embankment base	–	***	*	–
Column radial displacements	***	***	–	**
Pore pressures in soft soil	***	***	**	–
Time variations	***	***	**	–
Factor of safety against overall failure	–	***	**	–
General comments	Most common numerical analyses	More general and comprehensive results are provided; time consuming	Lacks additional validation studies	Straightforward analysis; to be performed before numerical analyses

*** Very applicable
** Applicable
* Less suitable
– Not suitable

advisable to be performed considering it is more time consuming. The 3D strip analysis provides all necessary information, but it may be justified just in special projects and is more commonly used in research studies. Any numerical analysis should be preceded by calculations using an analytical model, such as that proposed by Raithel and Kempfert (2000).

Annex I

Pre-design charts of Geosynthetic Encased Granular Columns (GECs) "$E_{oed,ref}$ = 500 kPa"

Group "A" charts

Soft clay strength properties: $c' = 2$ kPa and $\phi' = 25°$

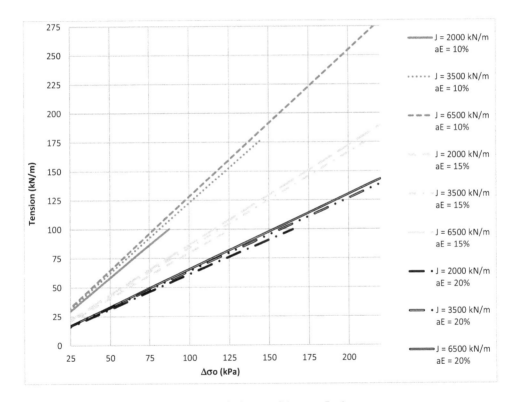

Chart IA Geosynthetic tensile force vs. applied stress ($H_{soft\ soil}$ = 5 m)

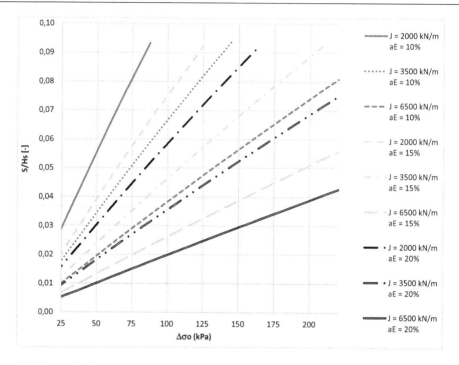

Chart 2A Normalized settlement *vs.* applied stress (H$_{soft\ soil}$ = 5 m)

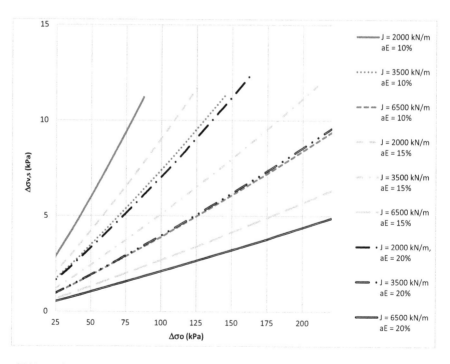

Chart 3A Vertical stress on soil *vs.* applied stress (H$_{soft\ soil}$ = 5 m)

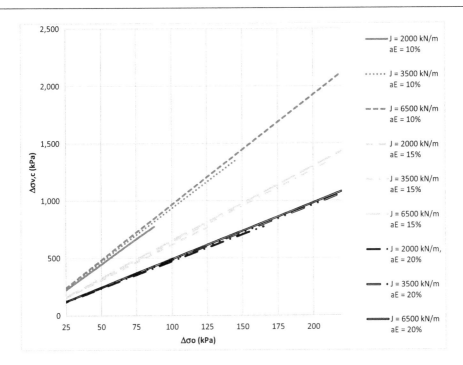

Chart 4A Vertical stress on column *vs.* applied stress (H$_{\text{soft soil}}$ = 5 m)

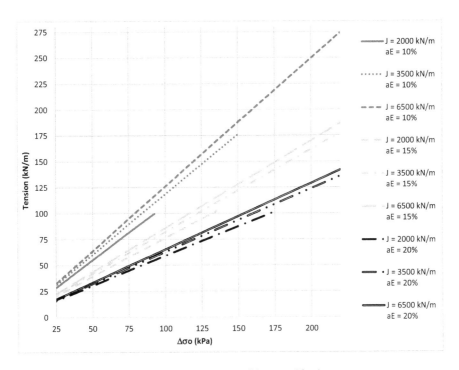

Chart 5A Geosynthetic tensile force *vs.* applied stress (H$_{\text{soft soil}}$ = 10 m)

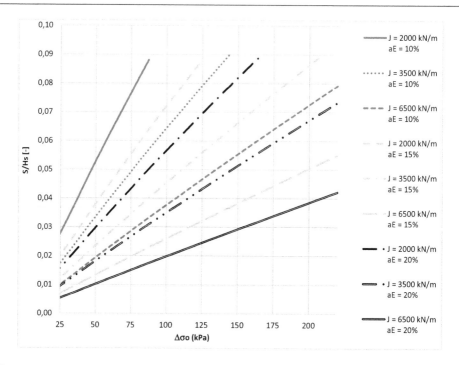

Chart 6A Normalized settlement *vs.* applied stress ($H_{soft\ soil}$ = 10 m)

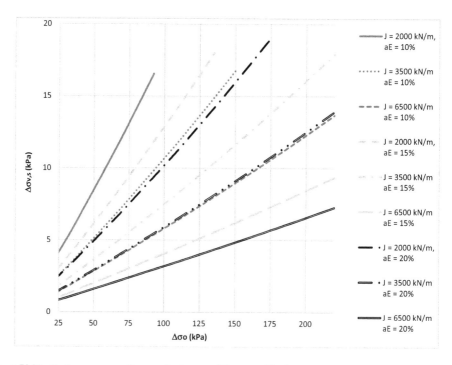

Chart 7A Vertical stress on soil *vs.* applied stress ($H_{soft\ soil}$ = 10 m)

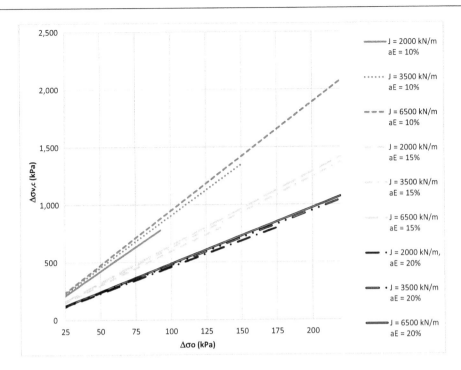

Chart 8A Vertical stress on column *vs.* applied stress ($H_{soft\ soil}$ = 10 m)

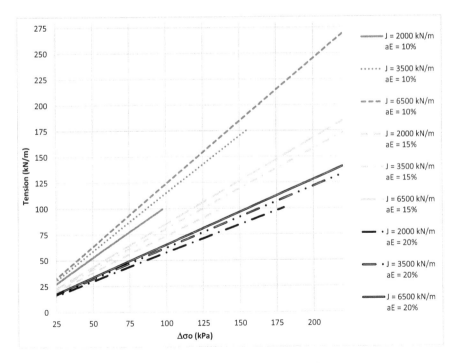

Chart 9A Geosynthetic tensile force *vs.* applied stress ($H_{soft\ soil}$ = 15 m)

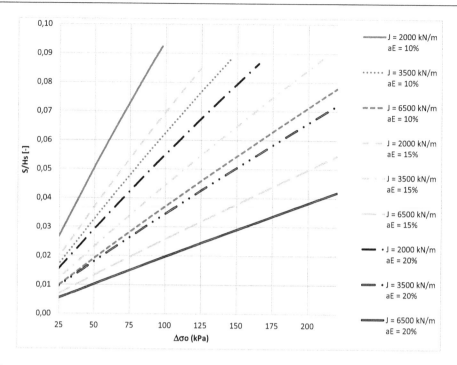

Chart 10A Normalized settlement *vs.* applied stress (H$_{soft soil}$ = 15 m)

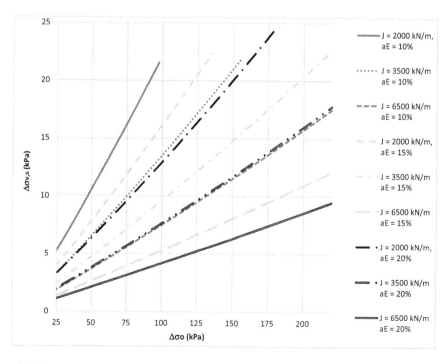

Chart 11A Vertical stress on soil *vs.* applied stress (H$_{soft soil}$ = 15 m)

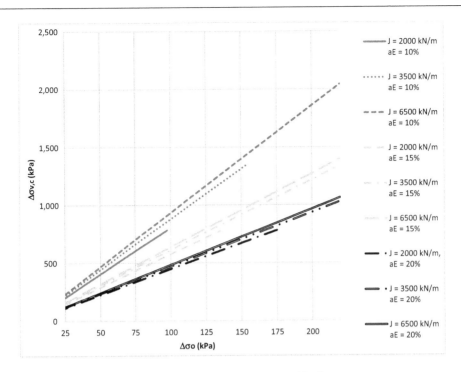

Chart 12A Vertical stress on column *vs.* applied stress (H$_{soft soil}$ = 15 m)

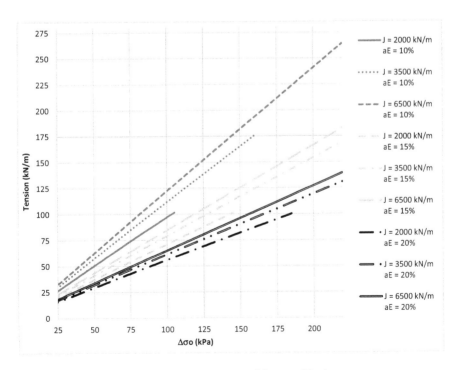

Chart 13A Geosynthetic tensile force *vs.* applied stress (H$_{soft soil}$ = 20 m)

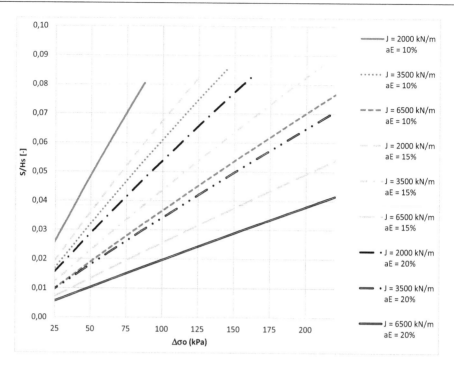

Chart 14A Normalized settlement *vs.* applied stress ($H_{soft\ soil}$ = 20 m)

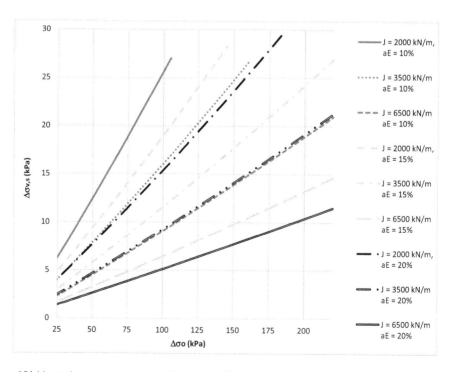

Chart 15A Vertical stress on soil *vs.* applied stress ($H_{soft\ soil}$ = 20 m)

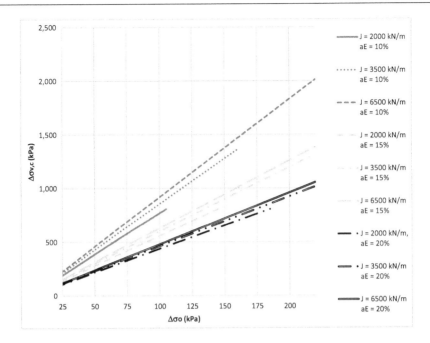

Chart 16A Vertical stress on column *vs.* applied stress ($H_{soft\ soil}$ = 20 m)

Group "B" charts

Soft clay strength properties: c'= 5 kPa and ϕ' = 28°

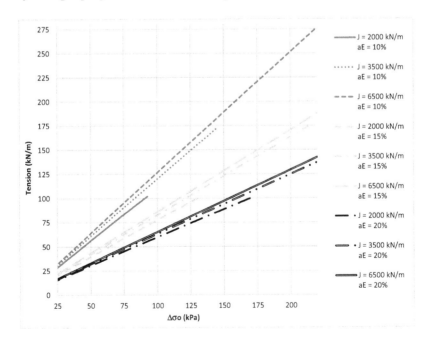

Chart 1B Geosynthetic tensile force *vs.* applied stress ($H_{soft\ soil}$ = 5 m)

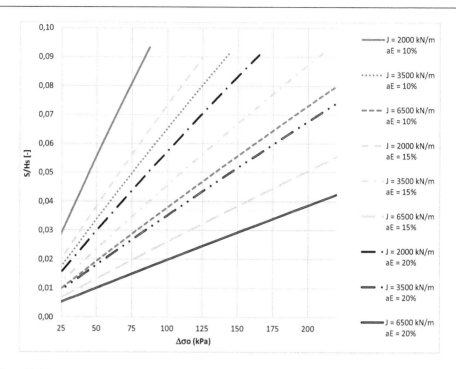

Chart 2B Normalized settlement *vs.* applied stress (H$_{soft\,soilol}$ = 5 m)

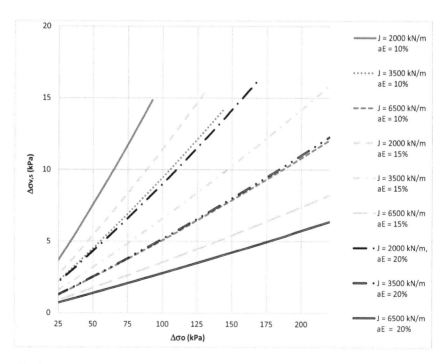

Chart 3B Vertical stress on soil *vs.* applied stress (H$_{soft\,soil}$ = 5 m)

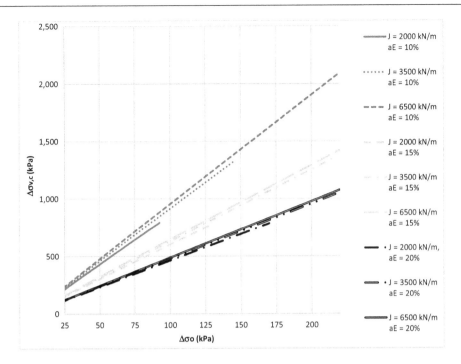

Chart 4B Vertical stress on column *vs.* applied stress (H$_{soft\ soil}$ = 5 m)

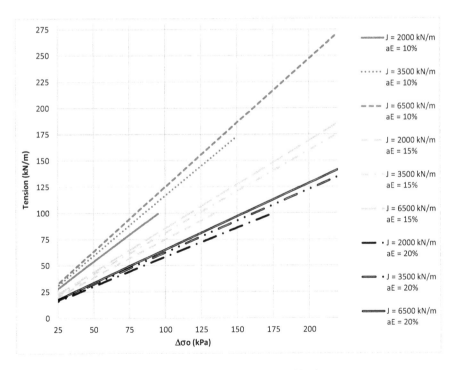

Chart 5B Geosynthetic tensile force *vs.* applied stress (H$_{soft\ soil}$ = 10 m)

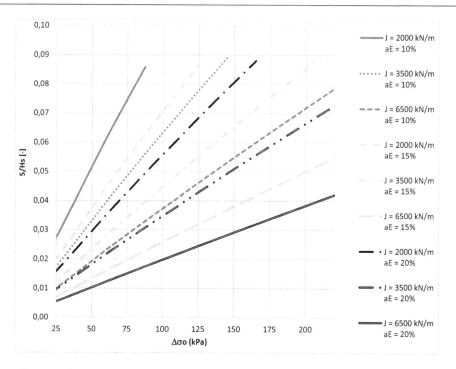

Chart 6B Normalized settlement *vs.* applied stress (H_soft soil = 10 m)

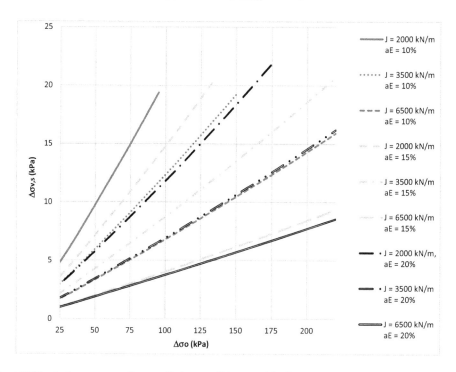

Chart 7B Vertical stress on soil *vs.* applied stress (H_soft soil = 10 m)

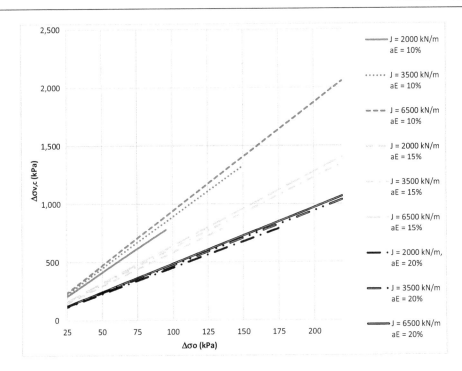

Chart 8B Vertical stress on column *vs.* applied stress ($H_{\text{soft soil}} = 10$ m)

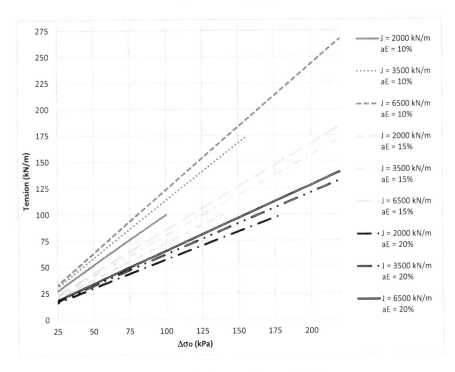

Chart 9B Geosynthetic tensile force *vs.* applied stress ($H_{\text{soft soil}} = 15$ m)

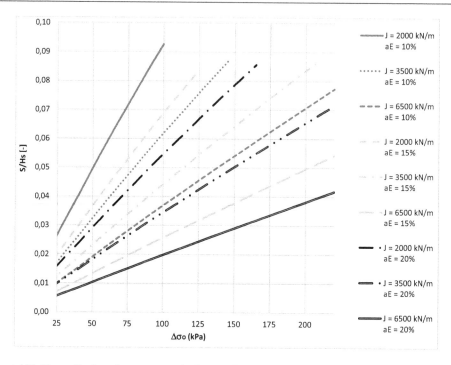

Chart 10B Normalized settlement *vs.* applied stress ($H_{soft\ soil}$ = 15 m)

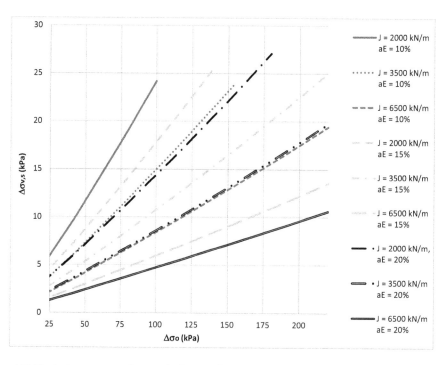

Chart 11B Vertical stress on soil *vs.* applied stress ($H_{soft\ soil}$ = 15 m)

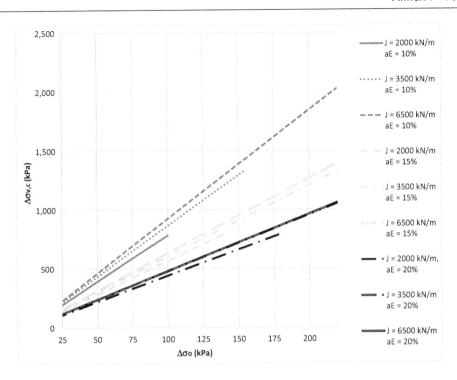

Chart 12B Vertical stress on column *vs.* applied stress (H$_{\text{soft soil}}$ = 15 m)

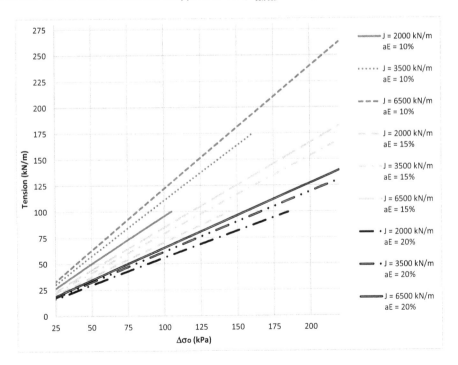

Chart 13B Geosynthetic tensile force *vs.* applied stress (H$_{\text{soft soil}}$ = 20 m)

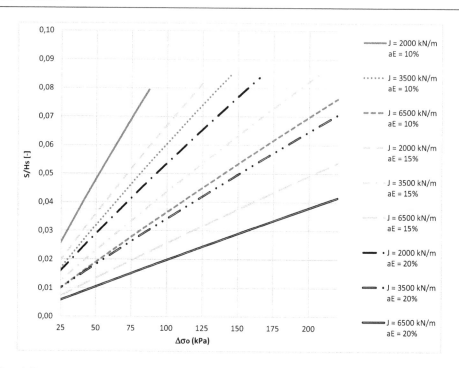

Chart 14B Normalized settlement *vs.* applied stress ($H_{soft\ soil}$ = 20 m)

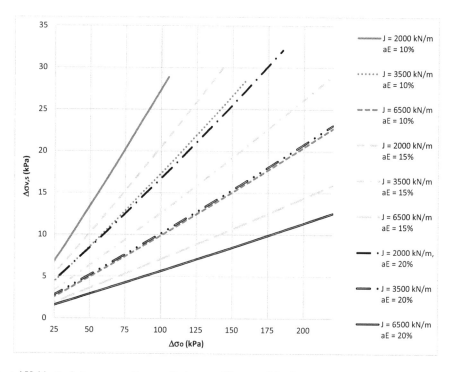

Chart 15B Vertical stress on soil *vs.* applied stress ($H_{soft\ soil}$ = 20 m)

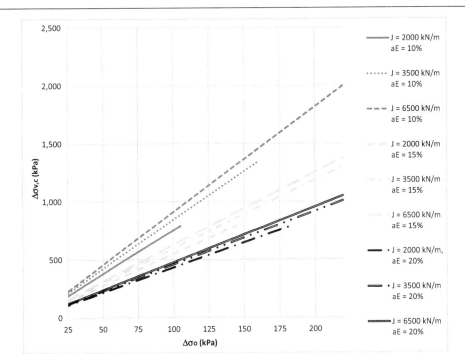

Chart 16B Vertical stress on column *vs.* applied stress (H_{soft soil} = 20 m)

Annex II

Pre-design charts of Geosynthetic
Encased Granular Columns (GECs)
"$E_{oed,ref}$ = 1500 kPa"

Group "C" charts

Soft clay strength properties: $c' = 2$ kPa and $\phi' = 25°$

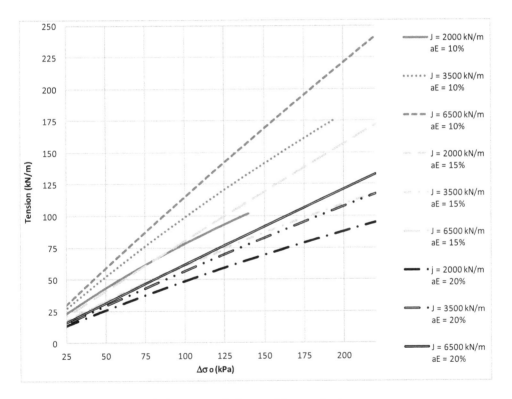

Chart IC Geosynthetic tensile force vs. applied stress ($H_{soft\ soil}$ = 5 m)

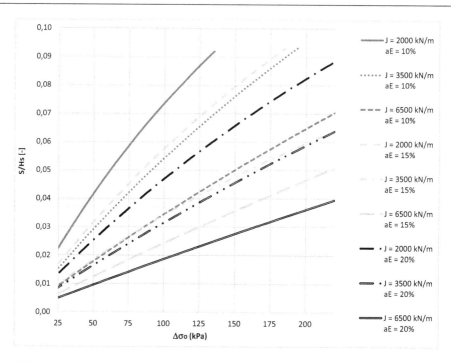

Chart 2C Normalized settlement *vs.* applied stress ($H_{\text{soft soil}}$ = 5 m)

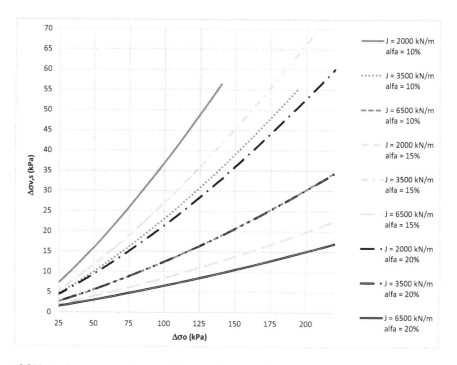

Chart 3C Vertical stress on soil *vs.* applied stress ($H_{\text{soft soil}}$ = 5 m)

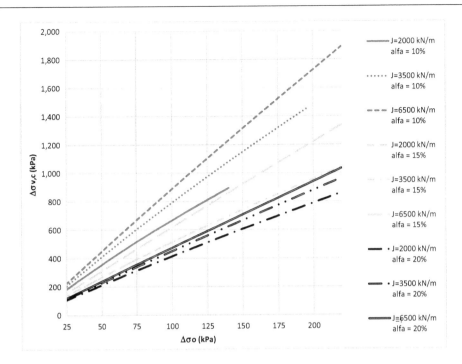

Chart 4C Vertical stress on column *vs.* applied stress (H$_{soft\ soil}$ = 5 m)

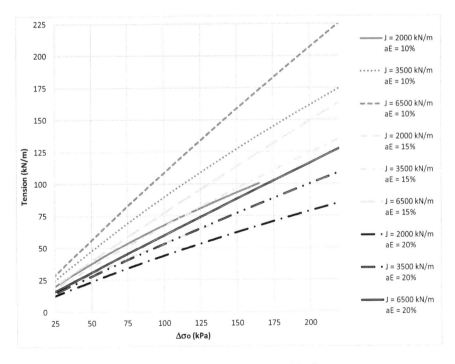

Chart 5C Geosynthetic tensile force *vs.* applied stress (H$_{sft\ soil}$ = 10 m)

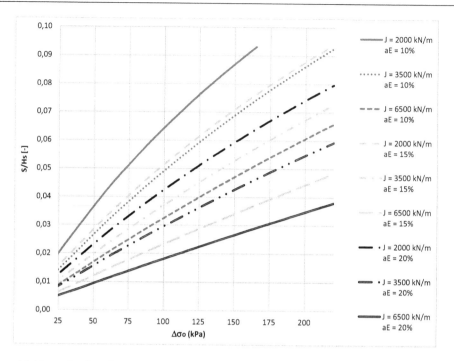

Chart 6C Normalized settlement *vs.* applied stress (H$_{\text{soft soil}}$ = 10 m)

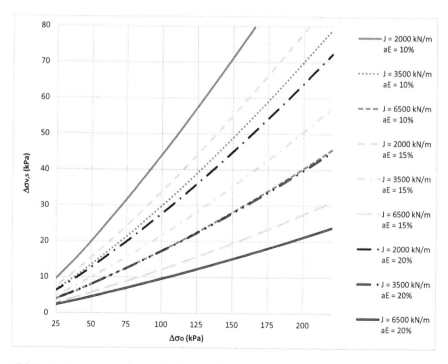

Chart 7C Vertical stress on soil *vs.* applied stress (H$_{\text{soft soil}}$ = 10 m)

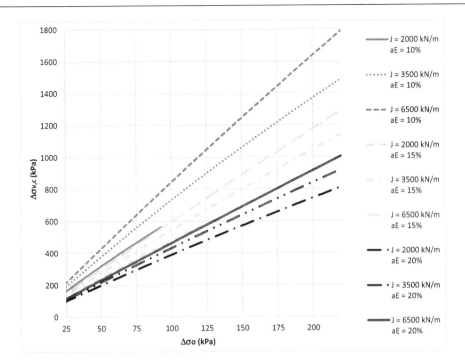

Chart 8C Vertical stress on column *vs.* applied stress (H$_{soft soil}$ = 10 m)

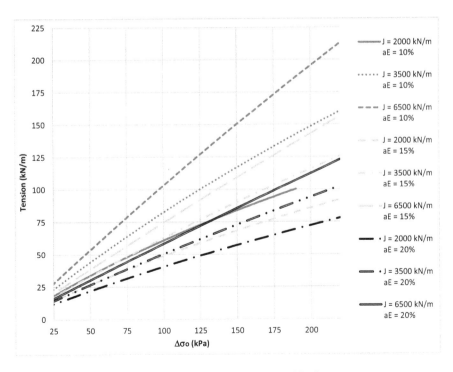

Chart 9C Geosynthetic tensile force *vs.* applied stress (H$_{soft soil}$ = 15 m)

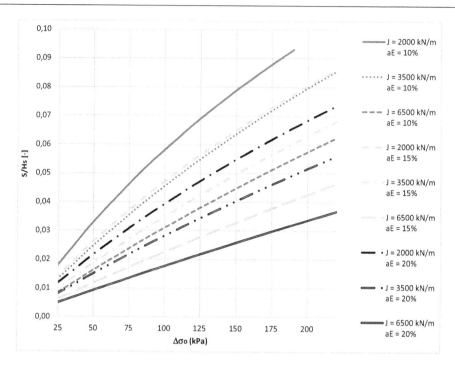

Chart 10C Normalized settlement *vs.* applied stress (H$_{soft soil}$ = 15 m)

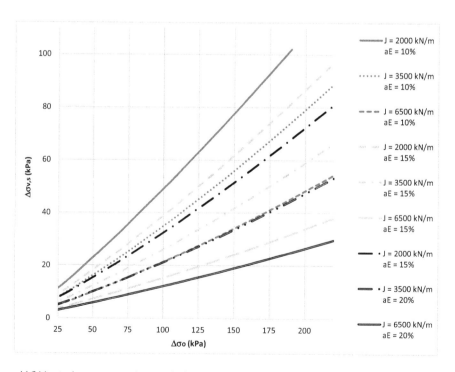

Chart 11C Vertical stress on soil *vs.* applied stress (H$_{soft soil}$ = 15 m)

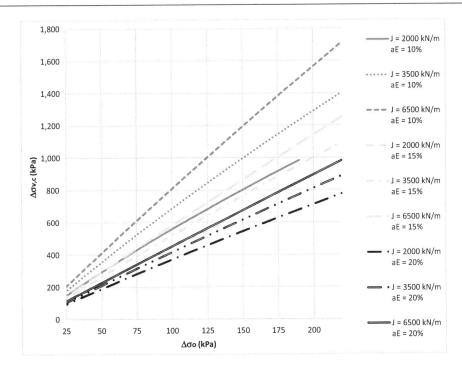

Chart 12C Vertical stress on column *vs.* applied stress (H$_{\text{soft soil}}$ = 15 m)

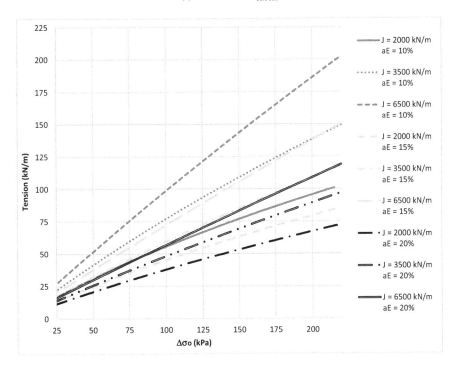

Chart 13C Geosynthetic tensile force *vs.* applied stress (H$_{\text{soft soil}}$ = 20 m)

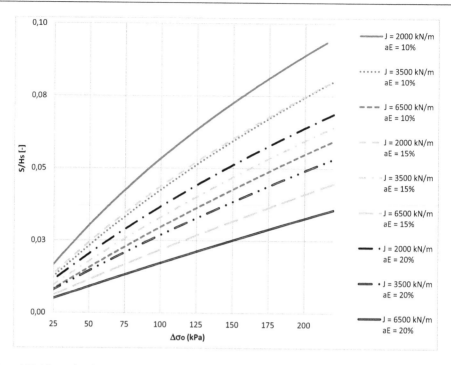

Chart 14C Normalized settlement *vs.* applied stress ($H_{soft\ soil}$ = 20 m)

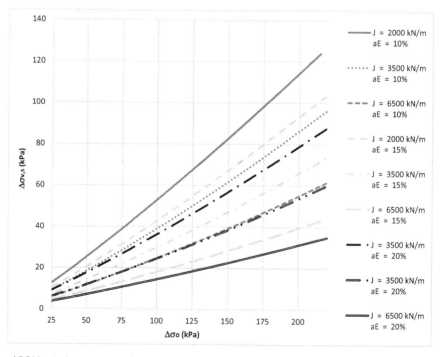

Chart 15C Vertical stress on soil *vs.* applied stress ($H_{soft\ soil}$ = 20 m)

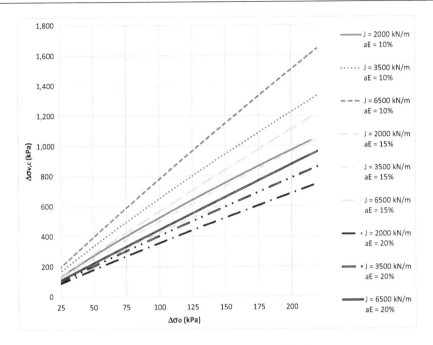

Chart 16C Vertical stress on column *vs.* applied stress ($H_{soft\ soil}$ = 20 m)

Group "D" charts

Soft clay strength properties: c′ = 5 kPa and φ′ = 28°

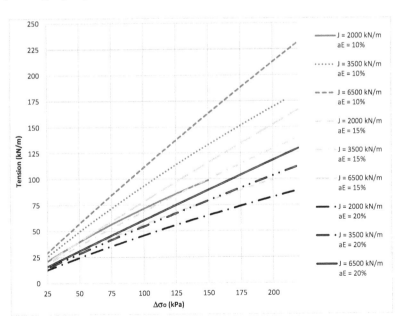

Chart 1D Geosynthetic tensile force *vs.* applied stress ($H_{soft\ soil}$ = 5 m)

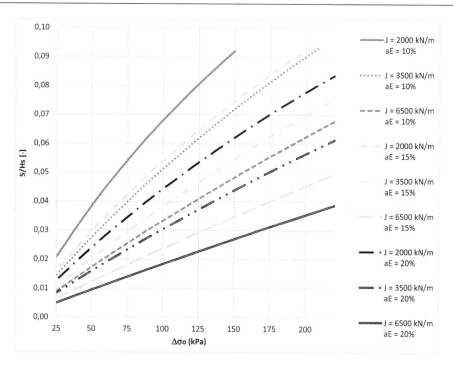

Chart 2D Normalized settlement *vs.* applied stress (H$_{soft\ soil}$ = 5 m)

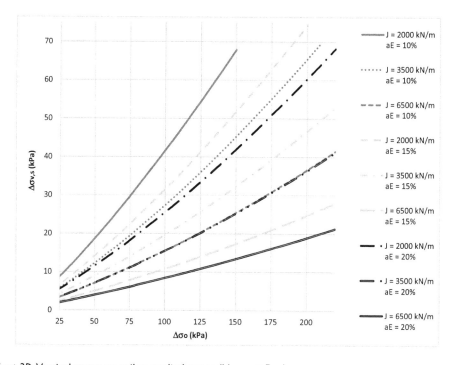

Chart 3D Vertical stress on soil *vs.* applied stress (H$_{soft\ soil}$ = 5 m)

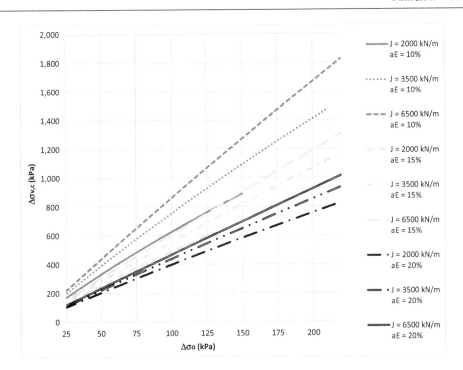

Chart 4D vertical stress on column vs. applied stress (H$_{\text{soft soil}}$ = 5 m)

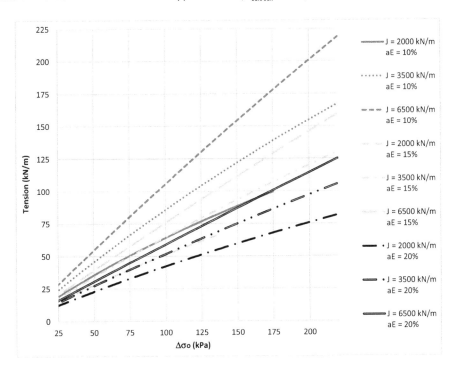

Chart 5D Geosynthetic tensile force vs. applied stress (H$_{\text{soft soil}}$ = 10 m)

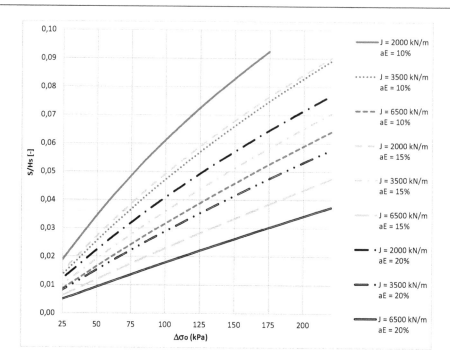

Chart 6D Normalized settlement *vs.* applied stress ($H_{soft\ soil}$ = 10 m)

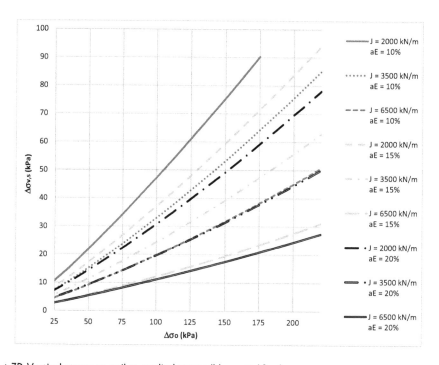

Chart 7D Vertical stress on soil *vs.* applied stress ($H_{soft\ soil}$ = 10 m)

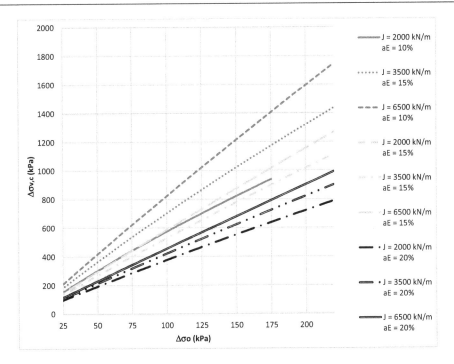

Chart 8D Vertical stress on column *vs.* applied stress ($H_{soft\ soil}$ = 10 m)

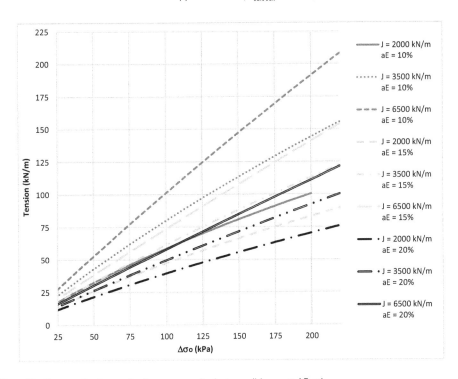

Chart 9D Geosynthetic tensile force *vs.* applied stress ($H_{soft\ soil}$ = 15 m)

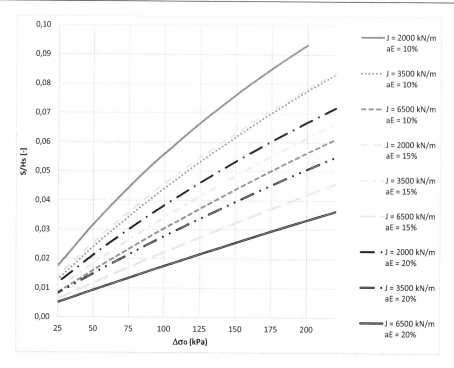

Chart 10D Normalized settlement *vs.* applied stress (H$_{\text{soft soil}}$ = 15 m)

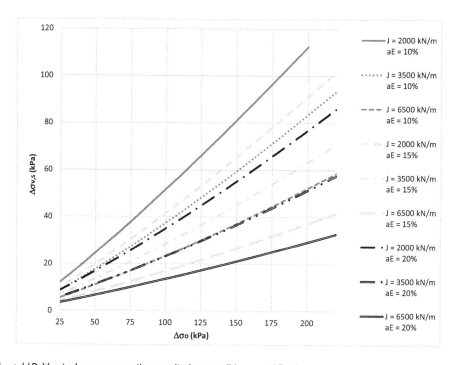

Chart 11D Vertical stress on soil *vs.* applied stress (H$_{\text{soft soil}}$ = 15 m)

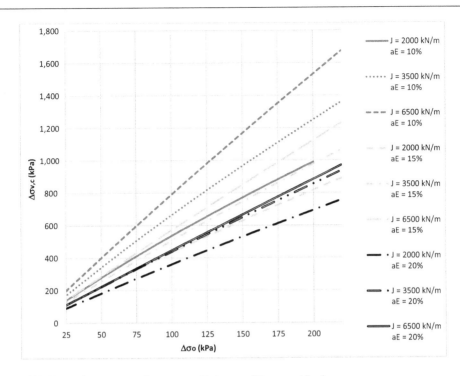

Chart 12D Vertical stress on column *vs.* applied stress (H$_{\text{soft soil}}$ = 15 m)

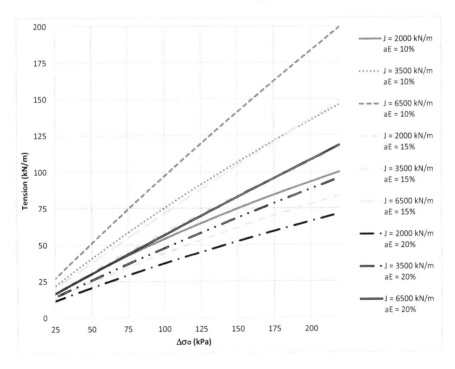

Chart 13D Geosynthetic tensile force *vs.* applied stress (H$_{\text{soft soil}}$ = 20 m)

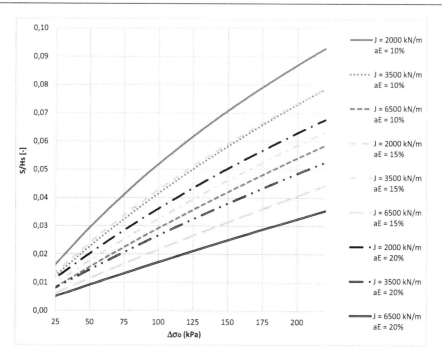

Chart 14D Normalized settlement *vs.* applied stress (H$_{soft soil}$ = 20 m)

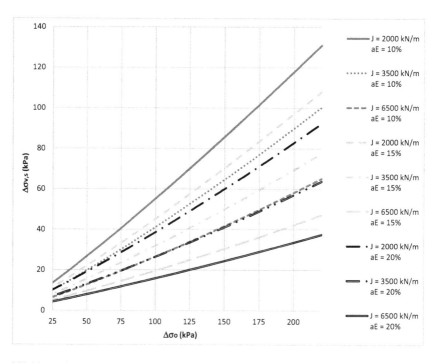

Chart 15D Vertical stress on soil *vs.* applied stress (H$_{soft soil}$ = 20 m)

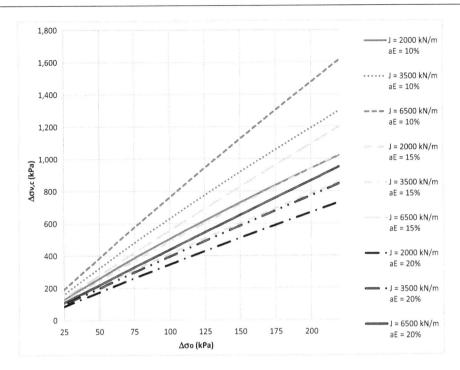

Chart 16D Vertical stress on column *vs.* applied stress ($H_{soft\ soil}$ = 20 m)

References

Aboshi, H., Ichimoto, E., Enoki, M. & Harada, K. (1979). The composer-a method to improve characteristics of soft clays by inclusion of larger diameter sand columns. *Proceeding of International Conference on Soil Reinforcement: Reinforced Earth and Other Techniques*, Paris, France, Volume 1, pp. 211–216.

Alexiew, D., Blume, K.-H. & Raithel, M. (2016). Bridge approach on Geosynthetic Encased Columns (GEC) in northern Germany: Measurement program and experience. *Proceedings GeoAmericas*, Miami Beach, FL, USA, pp. 378–387.

Alexiew, D., Brokemper, D. & Lothspeich, S. (2005). Geotextile Encased Columns (GEC): Load capacity, geotextile selection and pre-design graphs. *Proceedings of the Geo-Frontiers Conference*, Geotechnical Special Publication, Austin, TX, USA, pp. 497–510.

Alexiew, D., Horgan, G.J. & Brokemper, D. (2003). Geotextile encased columns (GEC): Load capacity and geotextile selection. *BGA International Conference on Foundation: Innovations, Observations, Design and Practice*, Dundee, Scotland, pp. 81–90.

Alexiew, D., Moormann, C. & Jud, H. (2010). Foundation of coal/coke stockyard on soft soil with geotextile encased columns and horizontal reinforcement. *Proceedings of the 9th International Conference on Geosynthetics*, Guarujá, Brazil, pp. 1905–1909.

Alexiew, D. & Raithel, M. (2015). Geotextile Encased Columns (GEC): Case studies over twenty years. In: Indraratna, B., Jian, C. & Rujikiatkamjorn, C. (eds) *Ground Improvement Case Histories: Embankments with Special Reference to Consolidation and Other Physical Methods*. Elsevier, Amsterdam, Netherlands.

Alexiew, D., Raithel, M., Küster, V. & Detert, O. (2012). 15 years of experience with geotextile encased granular columns as foundation system. *Proceedings International Symposium on Ground Improvement IS-GI, ISSMGE TC 211*, Brussels, CD, no pages.

Alexiew, D., Raithel, M., Schimmel, L. & Schröer, S. (2015). Geotextile Encased Columns (GEC) as pile-similar foundation elements: Basics, specifics, case studies. *40th Annual Conference on Deep Foundations*, Oakland, CA, USA, pp. 441–452.

Alexiew, D. & Thomson, G. (2013). Foundations on geotextile encased granular columns: Overview, experience, perspectives. *International Symposium on Advances in Foundation Engineering (ISAFE 2013)*, Singapore, pp. 401–407.

Alexiew, D. & Thomson, G. (2014). Geotextile Encased Columns (GEC): Why, where, when, what, how? *Fourth International Conference on Geotechnique, Construction Materials and Environment*, Brisbane, Australia, pp. 484–489.

Almeida, M.S.S., Britto, A.M. & Parry, R.H.G. (1986). Numerical modeling of a centrifuged embankment on soft clay. *Canadian Geotechnical Journal*, 23, 103–114.

Almeida, M.S.S., Hosseinpour, I. & Riccio, M. (2013). Performance of a Geosynthetic-Encased Column (GEC) in soft ground: Numerical and analytical studies. *Geosynthetics International*, 20(4), 252–262.

Almeida, M.S.S., Hosseinpour, I., Riccio, M. & Dimiter, A. (2014). Behavior of geotextile-encased granular columns supporting test embankment on soft deposit. *Journal of Geotechnical and Geoenvironmental Engineering*, 141(3), doi: 04014116-04014116-9.

Almeida, M.S.S. & Marques, M.E.S. (2013). *Design and Performance of Embankments on Very* Soft Soils. Taylor and Francis, CRC Press, London.

Ayadat, T. & Hanna, A.M. (2005). Encapsulated stone columns as a soil improvement technique for collapsible soil. *Ground Improvement*, 9(4), 137–147.

Balaam, N.P. & Booker, J.R. (1981). Analysis of rigid rafts supported by granular piles. *International Journal for Numerical Methods in Geomechanics*, 5, 379–403.

Barksdale, R.D. & Bachus, R.C. (1983). Design and construction of stone columns. Report FHWA/ RD-83/026, National Technical Information Service, Springfield, VA, USA.

Barron, R.A. (1948). Consolidation of fine-grained soils by drain wells. *Journal of the Soil Mechanics and Foundation Division, ASCE*, 73(6), 811–835.

Brinkgreve, R.B.J. & Vermeer, P.A. (2012). *PLAXIS: Finite Element Code for Soil and Rock Analyses*, Version 10. Balkema, Rotterdam, Netherlands.

BS 8006. (2010). *Code of Practice for Strengthened/Reinforced Soils and Other Fills*. BSI Standards Publication, London.

Carreira, M., Almeida, M.S.S. & Pinto, A. (2016). A numerical study on the critical height of embankments supported by geotextile encased granular columns. *Procedia Engineering*, 143, 1341–1349.

Castro, J. (2017). Modeling stone columns. *Materials*, 10, 782; MDPI, doi: 10.3390/ma10070782.

Castro, J. & Sagaseta, C. (2010). Deformation and consolidation around encased stone columns. *Geotextiles and Geomembranes*, 29, 268–276.

Castro, J. & Sagaseta, C. (2011). Deformation and consolidation around encased stone columns. *Geotextiles and Geomembranes*, 29(3), 268–276.

Castro, J. & Sagaseta, C. (2013). Influence of elastic strains during plastic deformation of encased stone columns. *Geotextiles and Geomembranes*, 37, 45–53.

Chen, J.F., Li, L.Y., Xue, J.F. & Feng, S.Z. (2015). Failure mechanism of geosynthetic-encased stone columns in soft soils under embankment. *Geotextiles & Geomembranes*, 43, 424–443.

De Beer, E.E. & Wallays, M. (1972). Forces induced in piles by unsymmetrical surcharges on the soil around the piles. *Proceedings of the 5th European CSMFE*, Madrid, Spain, Volume 1, pp. 325–332.

De Mello, L.G., Mandolfo, M., Montez, F., Tsukahara, C.N. & Bilfinger, W. (2008). First use of geosynthetic encased sand columns in South America. *1st Pan American Geosynthetics Conference & Exhibition*, CD-ROM, Cancun, Mexico.

Di Prisco, C., Galli, A., Cantarelli, E. & Bongiorno, D. (2006). Georeinforced sand columns: Small scale experimental tests and theoretical modeling. *Proceedings of the 8th International Conference on Geosynthetics*, Yokohama, Japan, pp. 1685–1688.

EBGEO. (2011). *Recommendations for Design and Analysis of Earth Structures Using Geosynthetic Reinforcements*, EBGEO (English version). German Geotechnical Society (DGGT), Ernst & Sohn, Essen-Berlin. 313 p.

Ghionna, V. & Jamiolkowski, M. (1981). Colonne di ghiaia, X Ciclo dedicato ai Problemi di Meccanica dei Terreni ed Ingegneria delle Fondazioni, Instituto Politécnico de Turim, Turim, Italy, Atti Istituto di Scienza e Tecnica delle Costruzioni del Politecnicodi Torino n. 507.

Gniel, J. & Bouazza, A. (2010). Construction of geogrid encased stone columns: A new proposal based on laboratory testing. *Geotextiles and Geomembranes*, 28(1), 108–118.

Greenwood, D.A. (1970). Mechanical improvement of soils below ground surface. *Proceeding of Ground Engineering Conference*, Institution of Civil Engineers, London, England, pp. 11–22.

Han, J. & Gabr, M.A. (2002). Numerical analysis of geosynthetic-reinforced and pile-supported earth platform over soft soil. *Geotechnical and Geoenvironmental Engineering, ASCE*, 128, 44–53.

Han, J. & Ye, S.L. (2002). A theoretical solution for consolidation rates of stone column-reinforced foundations accounting for smear and well resistance effects. *The International Journal of Geomechanics*, 2(2), 135–151.

Hatami, K. & Bathurst, R.J. (2005). Development and verification of a numerical model for the analysis of geosynthetic-reinforced soil segmental walls under working stress conditions. *Canadian Geotechnical Journal*, 42(4), 1066–1085.

Hosseinpour, I. (2015). *Test Embankment on Geotextile Encased Granular Columns Stabilized Soft Ground*. PhD thesis, Graduate School of Engineering COPPE, Federal University of Rio de Janeiro UFRG, Rio de Janeiro, Brazil.

Hosseinpour, I., Almeida, M.S.S. & Riccio, M. (2015). Full-scale load test and finite-element analysis of soft ground improved by geotextile-encased granular columns. *Geosynthetics International*, 22(6), 428–438.

Hosseinpour, I., Almeida, M.S.S. & Riccio, M. (2016). Ground improvement of soft soil by geotextile-encased columns. *Ground Improvement*, 169(4), 297–305.

Hosseinpour, I., Almeida, M.S.S., Riccio, M. & Baroni, M. (2017). Strength and compressibility characteristics of a soft clay subjected to ground treatment. *Geotechnical and Geological Engineering*, 35(3), 1051–1066.

Hosseinpour, I., Riccio, M. & Almeida, M.S.S. (2014). Numerical evaluation of a granular column reinforced by geosynthetics using encasement and laminated disks. *Geotextiles and Geomembranes*, 42(4), 363–373.

Hosseinpour, I., Riccio, M. & Almeida, M.S.S. (2017). Verification of a plane strain model for the analysis of encased granular column. *Journal of GeoEngineering*, 12(4), 137–145.

Kempfert, H.G., Jaup, A. & Raithel, M. (1997). Interactive behavior of a flexible reinforced sand column foundation in soft soils. *ISSMGE, 14th International Conference on Soil Mechanics and Geotechnical Engineering*, Hamburg, Germany, pp. 1757–1760.

Keykhosropur, L., Soroush, A. & Imam, R. (2012). 3D numerical analyses of geosynthetic encased stone columns. *Geotextiles and Geomembranes*, 35, 61–68.

Khabbazian, M.S., Kaliakin, V.N. & Meehan, C.L. (2015). Column supported embankments with geosynthetic encased columns: Validity of the unit cell concept. *Geotechnical and Geological Engineering*, 33(3), 425–442.

Khabbazian, M.S., Meehan, C.L. & Kaliakin, V.N. (2010). Numerical study of effect of encasement on stone column performance. *GeoFlorida 2010: Advances in Analysis, Modeling and Design*, 184–196.

King, K.H., Das, B.M., Puri, V.K., Yen, S.C. & Cook, E.E. (1993). Strip foundation on sand underlain by soft clay with geogrid reinforcement. *Proceedings of the Third International Offshore and Polar Engineering Conference*, Singapore, Volume 1, pp. 517–521.

Liu, K.W., Rowe, R.K., Su, Q., Liu, B. & Yang, Z. (2017). Long-term reinforcement strains for column supported embankments with viscous reinforcement by FEM. *Geotextiles & Geomembranes*, 45, 307–319.

Lo, S.R., Zhang, R. & Mak, J. (2010). Geosynthetic-encased stone columns in soft clay: A numerical study. *Geotextiles & Geomembranes*, 28, 292–302.

Lunne, T., Robertson, P.K. & Powell, J.J.M. (1997). *Cone Penetration Testing in Geotechnical Practice*. Blackie Academic & Professional, London.

Magnani, H.O. (2006). *Behavior of Test Embankment on Soft Clay Taken to the Failure*. Ph.D. Thesis, Graduate School of Engineering COPPE, Universidade Federal do Rio de Janeiro UFRJ, RJ, Brazil.

Magnani, H.O., Almeida, M.S.S. & Ehrlich, M. (2009). Behavior of two reinforced test embankments on soft clay. *Geosynthetics International*, 16(3), 127–138.

Malarvizhi, S.N. & Ilamparuthi, K. (2007). Comparative study on the behavior of encased stone column and conventional stone column. *Soils and Foundations*, 47(5), 873–885.

Malarvizhi, S.N. & Ilamparuthi, K. (2008). Numerical analysis of encapsulated stone columns. *12th International Conference of International Association for Computer Methods and Advances in Geomechanics (IACMAG)*, Goa, India, pp. 3719–3726.

McGuire, M., Sloan, J., Collin, J. & Filz, G. (2012). Critical height of column-supported embankments from bench-scale and field-scale tests. *ISSMGE-TC 211 International Symposium on Ground Improvement IS-GI Brussels*, Brussels, Belgium.

Mckenna, J.M., Eyre, W.A. & Wolstenholme, D.R. (1975). Performance of an embankment supported by stone columns in soft ground. *Geotechnique*, 25(1), 51–59.

Mohapatra, S.R., Rajargopal, K. & Sharma, J. (2017). 3-Dimensional numerical modeling of geosynthetic-encased granular columns. *Geotextiles and Geomembranes*, 45(3), 131–141.

Murugesan, S. & Rajagopal, K. (2006). Geosynthetic-encased stone column: Numerical evaluation. *Geotextile and Geomembranes*, 24(6), 349–358.

Murugesan, S. & Rajagopal, K. (2007). Model tests on geosynthetic-encased stone columns. *Geosynthetic International*, 11(6), 346–354.

Poorooshasb, H.B. & Meyerhof, G.G. (1997). Analysis of behavior of stone columns and lime columns. *Computers and Geotechnics*, 20(1), 47–70.

Pulko, B., Majes, B. & Logar, J. (2011). Geosynthetic-encased stone columns: Analytical calculation model. *Geotextiles and Geomembranes*, 29, 29–39.

Raithel, M. (1999). Zum Trag-und Verformungsverhalten von geokunststoffummantelten Sandsäulen. Schriftenreihe Geotechnik, Universität Gesamthochschule Kassel, Heft 6.

Raithel, M., Alexiew, D. & Küster, V. (2012). Loading test on a group of geotextile encased columns and analysis of the bearing and deformation behavior and global stability. *International Conference on Ground Improvement and Ground Control (ICGI 2012)*, University of Wollongong, Australia, pp. 703–708.

Raithel, M. & Henne, J. (2000). Design and numerical calculations of a dam foundation with geotextile coated sand columns. *Proceedings of the 4th International Conference on Ground Improvement Geosystems*, Helsinki, Finland, pp. 1–8.

Raithel, M. & Kempfert, H.G. (1999). Bemessung von geokunststoffummantelten Sandsäulen. *Die Bautechnik*, 76(12).

Raithel, M. & Kempfert, H.G. (2000). Calculations models for dam foundations with geotextile coated sand columns. *Proceedings of the International Conference on Geotechnical & Geological Engineering*, GeoEngg, Melbourne, Australia, p. 347.

Raithel, M., Kempfert, H.G. & Kirchner, A. (2002). Geotextile Encased Columns (GEC) for foundation of a dike on very soft soils. *Proceeding of 7th International Conference on Geosynthetics, Nice*, France, Balkma, pp. 1025–1028.

Raithel, M., Kirchner, A., Schade, C. & Leusink, E. (2005). Foundation of constructions very soft soils with geotextile encased columns-state of the art. *Geotechnical Special Publication, Geo-Frontiers*, 1867–1877.

Riccio, M., Almeida, M.S.S. & Hosseinpour, I. (2012). Comparison of analytical and numerical methods for the design of embankments on geosynthetic encased columns. *Second Pan American Geosynthetics Conference and Exhibition*, GeoAmericas, Lima, Peru.

Riccio, M., Almeida, M.S.S., Rigolon, J.M. & Hosseinpour, I. (2016). Studies on soft soil treatment with Granular Encased Columns (GEC). *Event COPPEGEO: Symposium Willy Lacerda and Jacques de Medina, March*, Rio de Janeiro, Brazil.

Robertson, P.K. & Cabal, K.L. (2015). *Guide to Cone Penetration Testing for Geotechnical Engineering*, 6th edition. Gregg Drilling & Testing Inc., Signal Hill, California.

Schnaid, F. (2005). Geo-characterization and properties of natural soils by in situ tests. *Proceeding of 16th International Conference on Soil Mechanics and Geotechnical Engineering*, Osaka, Japan, Volume 1, pp. 3–46.

Schnaid, F., Winter, D., Silva, A.E.F., Alexiew, D. & Küster, V. (2017). Geotextile Encased Columns (GEC) used as pressure-relief system: Instrumented bridge abutment case study on soft soil. *Geotextiles and Geomembranes*, 45(3), 227–236.

Soderman, K.L. & Giroud, J.P. (1995). Relationships between uniaxial and biaxial stresses and strains in geosynthetics. *Geosynthetics Internationals*, 2(2), 495–504.

Tan, S.A., Tjahyono, S. & Oo, K.K. (2008). Simplified plane-strain modeling of stone-columns reinforced ground. *Geotechnical and Geoenvironmental Engineering, ASCE*, 134(2), 185–194.

Tandel, Y.K., Solanki, C.H. & Desai, A.K. (2012). Reinforced stone column: Remedial of ordinary stone column. *International Journal of Advances in Engineering & Technology*, 3(2), 340–348.

Tavenas, F., Mieussens, C. & Bourges, F. (1979). Lateral displacement in clay foundations under embankments. *Canadian Geotechnical Journal*, 16(3), 532–550.

Tschebotarioff, G.P. (1962). *Retaining Structures, Chapter 5 in Foundation Engineering*. G.A. Leonards (ed). McGraw-Hill Book Co., New York, p. 493.

Tschebotarioff, G.P. (1970). Bridge abutments on piles driven through plastic clay. *Proceedings Conference: Design and Installation of Pile Foundation and Cellular Structures*, Lehigh University, Pennsylvania, pp. 225–238.

Tschebotarioff, G.P. (1973). *Foundations, Retaining and Earth Structures*, 2nd edition. McGraw-Hill Kogakusha LTDA, Tokyo, Japan.

Van Impe, W. & Silence, P. (1986). Improving of bearing capacity of weak hydraulic fills by means of geotextiles. *Proceedings of the 3rd International Conference on Geotextiles*, Vienna, Austria, pp. 1411–1416.

Wehr, J. (2006). The undrained cohesion of the soil as a criterion for the column installation with a depth vibrator. *Proceedings of TransVib Conference*, Paris, France, pp. 157–162.

Yee, Y.W. & Raju, V.R. (2007). Ground improvement using vibro-replacement (vibro stone columns): Historical development, advancements and case histories in Malaysia. *16th Southeast Asian Geotechnical Conference*, Kuala Lumpur, Malaysia.

Yoo, C. (2010). Performance of geosynthetic-encased stone column in embankment construction: Numerical investigation. *Geotechnical and Geoenvironmental Engineering, ASCE*, 136(8), 1148–1160.

Yoo, C. (2010). Performance of geosynthetic-encased stone columns in embankment construction: Numerical investigation. *Geotechnical and Geoenvironmental Engineering*, 136(8), 1148–1160.

Yoo, C. & Kim, S.B. (2009). Numerical modeling of geosynthetic-encased stone column-reinforced ground. *Geosynthetics International*, 16(3), 116–126.

Zhang, L. & Zhao, M. (2015). Deformation analysis of geotextile-encased stone columns. *The International Journal of Geomechanics*, 13(2), 04014053-1–04014053-10. doi: 10.1061/(ASCE)GM.1943-5622.0000389,04014053.

Zhang, N., Shen, S.L., Wu, H.N., Chai, J.C. & Xu, Y.S. (2015). Evaluation of effect of basal geotextile reinforcement under embankment loading on soft marine deposits. *Geotextiles and Geomembranes*, 43, 506–514.

Zhang, Y., Li, T. & Wang, Y. (2011). Theoretical elastic solutions for foundations improved by geosynthetic-encased columns. *Geosynthetics International*, 18(1), 12–20.

Author index

Subject index

Note: Page numbers in italic indicate a figure and page numbers in bold indicate a table on the corresponding page.

Milton Keynes UK
Ingram Content Group UK Ltd.
UKHW051849071024
449327UK00025B/1898